Sebastian Kneipp

The Codicil to My Will for the Healthy and the Sick

Sebastian Kneipp

The Codicil to My Will for the Healthy and the Sick

ISBN/EAN: 9783744783521

Printed in Europe, USA, Canada, Australia, Japan

Cover: Foto ©berggeist007 / pixelio.de

More available books at **www.hansebooks.com**

THE CODICIL

TO

"MY WILL"

FOR

THE HEALTHY AND THE SICK

CONTAINING

CHAPTERS ON THE ANATOMY AND CARE OF THE HUMAN BODY,
GYMNASTIC EXERCISES, FIRST HELP IN ACCIDENTS, COOKING
RECIPES, MEDICINAL PLANTS AND THE CURE OF DISEASES

BY

Mgr. SEBASTIAN KNEIPP,

PRIVY CHAMBERLAIN TO THE POPE AND PASTOR OF WOERISHOFEN.

KEMPTEN (BAVARIA).

JOS. KŒSEL PUBLISHER.

MDCCCXCVII.

PRINTED BY JOS. KŒSEL AT KEMPTEN (BAVARIA).

Pr̃eface.

y Will" of which this is a Supplement has followed the course of my earlier works viz. "My Water Cure" and "Thus shalt thou live" in becoming widely known and passing through many Editions.

This book which I have but just finished is a continuation of or second portion of "My Will" which explains why I have called it **"Codicil to My Will"**.

It often occurs in every day life that after a man has made his Will circumstances arise which render it necessary for him to make some additions to it; this is the case with me.

In this Codicil I am desirous of stating certain matter which is important and necessary to be known, and which was not treated of in the former works.

Should it be thought that this Codicil is somewhat late in appearing it must be remembered that my time for writing is necessarily very limited,

I think that the two parts of my Will, viz. the Will Proper and the Codicil will give full satisfaction as to my methods.

For greater completeness the second part of the Codicil contains a series of instructions and practical hints to the Laity so that they may be able to give assistance in cases of accidents and in the various other circumstances which threaten human life.

The Codicil gives instruction in Anatomy but in such a way that ordinary people can understand it. It is entirely for the laity and no previous anatomical knowledge is necessary to comprehend it.

I have thought it well to add some Gymnastic Exercises, which may easily be carried out after the water applications and which serve to make the joints more supple and the muscles stronger.

I give no preference to exercises requiring art and skill which are difficult to carry out and only recommend such movements as will be of use to the patient; for my books are all written in the first place for **sick** people even if in some of them, for example "Thus shalt thou live" I give hints and advice to the **healthy** how to live and brace themselves.

The Chapters on Anatomy, Gymnastics and First Aid to the Injured are written at my desire by my Secretary Herr Prior Reille, and the illustrations in the Anatomical portion of the book are copied with the greatest care from the Work of Dr. Joseph Ranke (Der Mensch) by the kind permission of the Publishers, (the Bibliographical Institute in Leipzig) and certainly these, together with the other numerous illustrations scattered through the book must, as I hope, greatly enrich this Codicil.

Thus, armed and equipped, 1 send forth this Codicil into the World for the use of suffering humanity who are seeking help, and for the instruction of all who read it.

Notwithstanding the many who recognise water as a curative medium yet the number of those who dispute its healing power is also very large.

As there is no universal remedy for death neither is there a cure for all sicknesses whether medicine or water be tried.

Proof however is at hand that one can heal by water and that cases of sickness have been cured by water in which neither medicine nor operative agencies had been successful.

One so readily turns up the nose at that which one does not know and has not tried, indeed I myself might do it had I not tried the good effect of water on my own body.

Still prejudices are sometimes overcome and I have the comfort of knowing that a large number of people have been convinced of the immense curative power of water even though at first they did not speak particularly respectfully of the water applications.

As in the preface to "My Will" I specially mentioned the company of Doctors who are disciples of my method and expressed a hope that they might

be always united and help to preserve my method
in its simplicity and purity, so here I gladly recognise the various Associations who are in my favour
and wish them well.

It in these Associations a naturally moderate
way of living, training and bracing is recommended,
and the Water Cure, with its manner of application
and effects explained, then their members and families will without difficulty be able to combat simple
diseases, and when the Doctor arrives he will find
intelligence and comprehension to carry out his orders without which the best prescriptions, if they
have to be carried out at home, lose much of their
good effect or may become useless.

If the Doctors and Associations find themselves
in accord and mutually support each other then both
sides will experience the greatest success from my
method.

This is my heartfelt wish.

Wörishofen, November 1896.

The Author.

THE CODICIL.

First Part.

General.

The young and the old
Who come to Herr Kneipp
Soon become healthy, agile and bright.

First Section.

A Word to the Sick.

Now that at length the second part of "My Will" is finished I add this Volume to that, just as one couples a railway carriage to a locomotive.

This part principally treats of illnesses which are described in so simple a manner that every one may understand.

What it means to be ill, only he can feel who has lost health and has for a long time been sick and miserable.

In such a case it is a kind and good action, no matter by whom performed to bring comfort and help to him.

The patient should on his part put entire trust in Him in whose hand rest the fate of men, their health, their sickness and all means by which lost health may be regained.

The longer I have to do with sick people, the more convinced I am that God, the Creator of all things, has bestowed upon us the means of Cure, viz. one half water the other half herbs.

Just as water, rightly applied, is the most harmless curative agent, so specially chosen herbs have great healing

1*

power: of the truth of this I am fully convinced by the countless results witnessed by me during many years.

Seeing that the Medical Profession boasts of so many learned men and that knowledge accomplishes so much in the preparation of curative remedies, it stands to reason that people would not come to the old Pastor, if it were not that so many who had been given up by the Doctors as incurable, have become well by the use of water under his care.

Still every patient should commence with prudence and use water and herbs in the simplest fashion.

The patient must remember that an evil just started may quickly be overcome, while an old trouble, which has taken deep hold, requires a certain amount of time to heal.

Many persons have come to me who have been ill for twenty or twenty-five years: these express their readiness to exercise any amount of patience, if only an improvement in their condition may be hoped for, never dreaming that they could enjoy perfect health again; and yet some of these have been completely restored.

There is a Seed Merchant living in my neighbourhood who for sixteen years sought help far and wide for a mitigation of his sufferings, but without avail. Yet at the present time, by means of water and herbs he may be considered one of the healthiest of men.

Therefore you who are sick have patience and do not look upon disease merely as an instrument for the chastisement of men, but rather as a means of acquiring eternal happiness.

As Curator of souls I have often thanked God that He has permitted diseases by means of which men are brought home and led out of evil ways.

How many who have been obliged to lie long in illness have seen the mistakes in their past lives, and have come to the conviction that all striving and struggling after earthly possessions is but vanity if care for the higher calling is neglected.

It often occurs that long illnesses bring about a complete change in men, so that many who have suffered thank God for the illness which brought them nearer to Him and taught them to try and be worthy of their eternal inheritance.

Every sick person should ask himself "Who has afflicted me with this illness?" "What is the intention of my heavenly Father with regard to it?"

The right answer will certainly not be wanting; and the pain and suffering will be like the bitter medicine which is indeed unpleasant to take, but which will restore us to health.

If water and herbs are to be the principal remedies, it is natural that a reasonable amount of care and attention should be bestowed upon them.

One of the most important things to secure both for the healthy and the sick is fresh, pure air.

There should be no sick-room without at least one little opening through which pure air may pass, and this is imperative where the patient suffers from unpleasant perspirations which are not only offensive to the nose, but are very likely to harm the healthy also.

Another necessity is **perfect cleanliness.**

It must not be forgotten that from every object an evaporation comes forth like a faint breath: Thus from the patient himself as from the sick-bed an evaporation issues which the nurse or attendant must breathe in or absorb.

Cleanliness relates to ablutions of the body, washing of the bed as well as of everything to be found in the sick-room.

A nurse who has known by personal experience what it is to be ill, can sympathize more with the patient, and one who has not known sickness, must be taught how the patient should be treated.

It is a very great mistake and one most unjurious to the patient to admit relations, acquaintances and neighbours to the sick-room.

It frequently happens that during such visits discussions arise about household matters, society, or politics and the poor invalid is confused and worried by all this talk and his condition considerably altered for the worse.

The patient requires his doctor, a quiet and patient nurse and a clergyman, otherwise he should see but few people or better still none at all.

Past and Present.

In human life important changes are always occurring; look at any trade or profession you please and you will note changes, nothing but changes!

Formerly thousands of people earned their daily bread in winter at the loom: this is no longer the case, for most articles now-a-days are made in factories.

Again, think how much used to be done at the lock-smith's workshop — now the smith does not even care to make a lock, for he can buy it at much less cost to himself.

Home hand-work at the present time is not of high character, as all able workers are employed in the factories.

Just as handicraft has undergone an important change, so also has medicine.

Sixty years ago there were still so called country doctors who cured principally with herbs and who employed also many laxatives and emetics.

I have known people who had to take a remedy every four weeks to prevent or cure headache and other evils.

At that time blood-letting was the fashion, the opinion being that disorders of the system and in the blood could only be removed by opening the veins. There were few diseases in which blood-letting was not resorted to.

If a woman was attacked by spasms, she was bled. Thus I knew a nun who had undergone blood-letting 300 times! The result one may easily imagine.

Another way of cure was by **dry-cupping** which was performed several times in the year. It was not uncommon to meet people, country ones especially, whose backs bore the marks of from fifteen to twenty cupping glasses.

Under such treatment it is no wonder that many diseases gained entrance to the system and that regular circulation of the blood was rendered impossible.

How much human blood the leech has absorbed! One cannot be surprised that we have so many anæmic people now. If people at that time had not kept themselves up by simple food and habits, much worse things would have happened.

Now-a-days these methods of cure are dispensed with or only resorted to in extreme cases.

It is quite certain that nature's laws having provided that no more blood can be formed in the body than the system requires, there can be no question of an overflow.

The regulation of the blood should always be a chief consideration and water is the only thing capable of undertaking this task.

In former days medicaments were prepared from harmless herbs; these in the present day have been mostly banished and their places taken by other preparations, minerals, etc., many of which unfortunately are poisonous.

Whether we are better off with these new remedies than we were formerly with the herbs we can ascertain by regarding closely the sick world of to-day and questioning old experienced people.

The herbs given us by God are as a rule crowded out by science. It has come to my knowledge over and over again that Doctors do not even know the names of many simple herbs much less their use! How unknown to many is pewter grass!

It is a matter much to be regretted that the use of poison is so common in advanced modern medicine. I cannot imagine upon what principle the modern science of healing works when in so many diseases it resorts at once to poisons. It is generally stated on that side that while poison in large quantities is harmful, in small quantities it is curative.

If one could go in and out among the sick who were being treated with poisons and even regard those who were cured by the same means and made inquiries of them what answer would one certainly get?

Dear Readers. We are all erring men! the more we study the less we know of God's beautiful World of nature! No mortal hand has yet created anything absolutely perfect: although the brain of man has found out much in the various branches of sciene. How far we have advanced with steam! What great and marvellous things have been brought about by Electricity of which formerly one had no conception!

What immense researches have been made into Chemistry and into the preparation and combination of various substances! How many important discoveries and inventions have come to light; while almost every profession has its new methods by which labour is perceptibly diminished. Medicine also has made great progress, whether to the advantage or prejudice of suffering humanity I will not decide, but this I feel bound to say that the progress thus made has resulted in moving us further and further from our former simple and natural means of healing.

To go back once more to the poisons — we will ask the patients what results have accrued to them from their use.

Quinine is, as a rule, the first remedy to which Doctors fly, especially in fevers; why not conclude the whole matter with a couple of washings? What is the effect of Quinine? Merely a momentary quieting.

If only every one who has to do with fever Patients would at least give cold water one trial in the manner I have indicated in the first part of "My Will", he could not help being convinced of the grand effect of Water!

Certainly, as I have already said, Quinine produces a momentary relief, but in a very short time the fever will show itself again more violently and not rarely in combination with other evils.

Therefore, dear people, who have to do with invalids, I commend it yet again to your hearts most emphatically: Spare your fellow creatures the use of poison, and you invalids beware of taking poisonous medicaments even though these remedies be pronounced excellent and even indispensable in certain cases: many invalids have come to the conviction that they are not indispensable.

The healthy and more emphatically the sick systems require simple, natural remedies, not poisons or artificial substances.

Morphia and Opium are again two of the poisons without which modern medicine would be helpless.

Are these really and truly curative remedies? No.

No one would think of indicating them as remedies, and as alleviating agencies they have little or no value.

It is true that injections of Morphia diminish pain very quickly and the patient feels relief for the time, but subsequently they produce the saddest results.

If only sick people would recognise in their sufferings a higher dispensation and not, in order to free themselves from pain, allow Doctors to stupify them with Morphia!

It is not right if man, finding himself in a condition ordained for a higher purpose, allows himself to be robbed of his sense and reason by Morphia in order to evade his affliction — such a man lies there like a brute beast if one may use the simile.

The system finds itself in a condition like that of tetanus.

Reason, sense and free will are taken away and the appearance of the man is as though he were hopelessly drunk. When after a time he comes to his senses the sufferings begin again and a second injection follows and then a third and so it goes on until the patient finally longs after it exactly as a thirsty person after a drink of water.

Then gradually a new evil arises, viz. **the craving for Morphia**, the results of which I have witnessed over and over again.

There are patients here constantly who have been completely ruined by this poison; the blood is poisoned, the stomach destroyed, and the brain so weakened that they are quite incapable of following any calling.

In their direst need they come to me and then, as you may suppose, Morphia is entirely forbidden them.

Therefore, dear readers, take no morphia, no opium, no strychnine in short no poisons whatever even in the very smallest doses.

Leave poison to those who desire to ruin their system by force and to those who still think to obtain good from it.

Practise instead a simple habit of life and do all that is included in that phrase and the result will be that many evils will disappear, and should you be visited with a really bad illness, think of Him who has afflicted you with it.

There is scarcely a house now-a-days in which coffee and tea are not found, whereas formerly coffee was scarcely known. It has now unfortunately spread to all circles and has established itself so firmly that people think they cannot live without it. The mother in her blindness gives her little one coffee and more often than not **bean** coffee: Because she herself drinks it with pleasure, she thinks it would be cruel to deny it to her darling.

But what does the mother look like? What indeed is the appearance of all coffee drinkers? They have pale and sallow complexions, while nervousness and chlorosis are constantly present.

Many may be inclined to ask "Why does coffee produce such bad results?" My answer is, "because it contains a poison which for want of a better name I call coffein".

If we compare the quiet simple life of sixty years ago with that of to-day, we cannot be surprised that even children are afflicted with all sorts of diseases formerly unknown but which are increasing year by year and making mankind miserable.

Sixty years ago housewives scarcely knew how to prepare coffee, while now it seems to be the only knowledge they can boast.

I am quite aware that all my preaching is for the most part useless; for although the truth of what I say is fully acknowledged, yet no one feels inclined to alter his manner of life, for his palate is spoiled and his whole system too much weakened to allow of change.

If you wish to be healthy, if you desire to be happy in your life and profession, listen to the words of an old priest who has nothing more at heart than the welfare of mankind and whose sincere desire it is to wean them from modern, ruinous habits and to lead them back to the happy paths of simplicity.

What is more beautiful than the love of simplicity! How happy and content people were formerly! And now?

Ah! everything is changed! one searches everywhere for all sorts of distractions; life is simply a struggle for pleasure and enjoyment and there are mistakes everywhere.

Formerly we were content with strengthening, natural whole-meal in our household; to-day only refined flour is used.

Many readers of my books, housewives, and especially corn merchants will ask "Why does the old Pastor lay such special stress on flour? Well, not without reason you may be sure: just eat a roll or cake made of coarse natural flour and then eat the same sort of thing, prepared with white refined flour and tell me which of the two is the better and the more strengthening.

Naturally the fastidious vitiated palates will prefer the latter, but all reasonable people will long after the former, because they know how strengthening and nourishing it is.

Many people eat only vegetable food; this is not sufficient where bread is made of refined flour; in this case meat must be eaten.

In Swabia flour food is universal, but everything is prepared with the natural flour.

It seems almost beyond belief that peasants can be so foolish as to carry their grain to the mills in order to exchange it for a white refined flour, yet many of them do so.

The bread made now-a-days by bakers is not to be compared with that of former times. It is all made now of refined flour and there is no such thing as wholesome, strengthening bread; moreover it is so light and small that one has to be careful it does not fly away.

Everything now is a matter of speculation, the desire being to obtain money and give nothing in exchange for it.

I wish there were more men like a country Magistrate I once knew:

If it came to this man's ears that a baker was selling bad bread or bread too light, he sent for a loaf and analysed it, and on finding it defective he caused the whole batch of bread baked that day to be taken from the shop and distributed among the poor: naturally the bakers made a note of this for their future guidance.

When the Magistrate saw that his method answered, he followed it up by advising the peasants to buy only of those bakers who sold good bread made of strengthening, natural flour and of proper weight.

He tried the same experiment with beer; unfortunately in this case he had to do with the owner of a brewery who was himself an exalted personage; he was offended and caused the universally beloved Magistrate to be removed to a lower position.

Be careful, dear housewives, to provide good bread for your families and if you find it impossible to get it at the bakers', make it yourselves according to the recipes set forth in my books.

Bread finds a place on every table and if it be not good, what else is there left to enjoy?

No less now-a-days are mistakes made as to clothing!

Formerly the proverb ran "Home spun, home made, is the most beautiful national costume", now fashion has ousted national costume everywhere. Modern town wear has now become almost universal.

Once the peasants and yokels wore good leather breeches which certainly were dear to buy, but on the other hand they lasted from ten to fifteen years.

I have myself worn such breeches made of buck skin and at the end of nine years they were in such good condition that I went on with them for another six.

If in course of time the breeches became somewhat shabby, one blackened them and again they became the most splendid breeches in the world.

Especially practical are these breeches for boys, for encased in them they can scramble about and fight as much as they please.

Now the fashionable garments that are just stuck together have penetrated even into the country. Vanity has gained the mastery and economy completely set aside.

Unluckily this is not merely the case among the higher classes but also among servants and artisans.

It has often happened to me that regarding a person approaching I have thought to myself "Here comes a noble countess" and after all it has been only a servant maid.

One day I drove from Buchloe to Turkheim and among the passengers was a very grandly dressed woman who, judging from her apparel, I thought must be at least a Baroness.

But soon I found out my mistake, for from her speech I learnt she was a servant going to a situation at Wörishofen. She complained to me that she had no more money and could not drive on to Wörishofen by the mail coach and went on to give me little veiled hints that I should pay for her. I would, however, on no account understand her and said to her "If you are able to get yourself up like a three-quarter countess you ought to have money enough to pay the coach fare." You see, dear Readers, how it is with this generation and it is not surprising.

The mothers themselves are not much better: "As the old ones sing so do the young ones twitter."

Still the blame is not always to be ascribed to the parents. Children who have been well educated go out into the world to earn their daily bread and it frequently happens that on receiving good wages they squander it recklessly on vanity and dissipation, and put by nothing for old age.

If only the heads of families would lay to heart the welfare of their children and dependants! If only they would strive to plant in the hearts of those entrusted to them by God the love of simplicity and content and extinguish as far as possible every germ of evil!

To do this successfully those in authority must set a good example. "Words move, example attracts by force." The Mother should never let her daughter play the fine lady and the Father should not make of his son a fool or a dandy. No matter what the rank or condition, the children should be constrained to work. A beautiful feather in the hat or kid gloves on the hand are no aids to work. "He who will not labour, neither shall he eat."

Place a peasant-maid or a sturdy country fellow who has always been accustomed to simplicity in dress and food, beside a town lady who is weakened and debilitated from head to foot, or next to a town dandy who looks in his fashionable clothes like a fool in his motley, and what a contrast you have!

Many, by means of my books, have been brought to see the mischief caused by this debilitating, foolish and vain life.

May all those who up to the present moment have been bound by fashionable and debilitating influences, at least make the attempt to brace their bodies. according to the instructions in "My Will" which are full and complete.

There are many families who have lived in constant discomfort and who have been under the care of Doctors from year's end to year's end, who would be well and happy if only they would brace their bodies and give up thinking of their supposed many infirmities.

Women suffer from the mistakes they make in dress, for it is well known that every folly which is exhibited in the fashionable papers, is copied by the modern women who never consider how seriously they injure their system by this folly.

We will take for example the corset:

If I were a law-giver I would levy a tax of fifty shillings on every pair of corsets. In this way the middle and lower classes would be spared injury for they could not afford the money. Let those who have the means quietly ruin their system if they please; they have after all nothing to do. I do not consider it beautiful for a woman to lace her body up until she looks like a greyhound.

Unluckily it has always been the fashion more or less to do so but not in so crass a manner as now: a supporting bodice or girdle used to be sufficient.

It is a crying evil that so many now-a-days act quite contrary to the law of nature. It is absolutely impossible that the use of corsets can do anything, except ruin the body.

A Doctor assured me that since the introduction of the wearing of corsets, the number of Doctors for diseases of the abdomen had also appreciably increased. Still people do not care about that so long as they can be in the fashion: if people thought at all they would be convinced that such a contraction of the body could result in nothing but harm.

We are taught this in the accidents and fatal cases resulting constantly from tight lacing.

Therefore, dear housewives, no corsets! They bring evil both to you and your posterity.

There is no shame in leaving the body as the good God gave it us; on the contrary, you sin deeply against God when you ruin your body by this senseless fashion.

The corset is not the only article injurious to the human body; the fashionable boot is almost equally so.

There is no surer way of ruining the foot than to force it into the now fashionable shoe, especially as it is now worn by women and even by men.

How is it we have so many diseases of the foot? Simply and solely from wearing the harmful fashionable shoe. It is quite impossible for the foot to develop encased in such narrow shoes which do not allow a proper circulation of the blood to take place.

Just wear simple suitable shoes in which no single part of the foot is unduly compressed.

For children the **soles of their feet** are the best shoes, and it seems to me very foolish for parents to put shoes and stockings on the feet of little children just out of long clothes.

I will take this opportunity of touching on another subject viz. the **elastic stocking.** This is very harmful to the foot, preventing its development and hindering the evaporation and circulation of the blood.

I will say no more; let those who believe me follow my advice, and those who do not must let it alone.

The fashions adopted by men are scarcely less foolish than those of women. All sorts of stupid articles are made now of which formerly the names were unknown. Breeches are worn which look like a sack while others are made so narrow that it is with difficulty they can be put off or on.

I should never finish if I were to relate all that has been conceived by the human mind in the way of foolish and harmful articles of clothing.

In spite of the fact that the modern world offers to supply our every want, we are not satisfied nor are we as happy as we used to be in the days gone by.

My opinion is that if we desire to be a really happy people we must go to school with our ancestors and let ourselves be instructed by them in simplicity, contentment and sufficiency.

Men must learn too that there is a higher happiness than that offered by modern life with its struggles and strivings which can only be obtained by religion.

As long as the higher happiness is neglected life will not be better with us but rather worse until we shall scarcely find a really happy person.

Partly the struggle after pleasure, partly debilitation and passions, and not least the carelessness about religion lead men into wrong paths and sometimes to destruction. Mankind is often to blame for its own misfortunes: if people had laid to heart the lessons learned from their ancestors the whole human family would stand on a higher footing.

Naturalness and simplicity are the great things to strive for.

As it is with the mode of life and articles of food so is it also with fluids. Everything is overrefined and poison which is now so often given to the system is injurious to it.

I will on no account praise myself but I must say that if every one would follow my example in simplicity of dress, food, manner of life and in bracing the system, a lusty and healthy race would result.

If I ever thought of giving up this simple mode of life and governed myself by the principles of modern fashion I should know that my days were numbered.

As it is I enjoy health in spite of my seventy-five years and I cannot thank the good God sufficiently that he has led me, who used to suffer so much, into this the only right way.

Go in for the Water Cure
Be attentive to Kneipp
And then you'll be sure
To be happy and bright

2

Something for Housewives.

From a good soil one obtains a good crop, but if the soil is poor one must not expect good fruit.

I may as well bring this simile forward while writing a few words for housewives.

No one should undertake matrimony, whether man or woman, if he or she be unsuited, that is if they be unsound, unhealthy and incapable of resisting the storms of life, otherwise they will be crushed under the burden they have undertaken.

The first requirement for married life is a good healthy development by which I mean a love of simplicity and of the simplest and most nourishing food, as well as of bracing.

If these virtues fail, very little good can be attained and matrimony may prove a double sorrow.

It may be that the man or woman belongs to a family in which consumption is inherent. I would say to such a one most earnestly "do not marry"; in any case he or she might live many years longer by remaining single. As it is with consumption so it is with other diseases and infirmities.

It is incumbent on every person who is thinking of marriage to ask "Have I the necessary qualifications for undertaking this burden?"

Not infrequently marriage leads to great possessions, fame, reputation or money and one thinks in thus marrying the person has won great happiness, forgetting to ask the important question "does the highest happiness of all exist in this union viz. health and religion?" As a rule these two last are not made much of and people plunge wildly and blindly into much misery.

I have stated my purpose of addressing a few words to housewives and I desire them to know that the following advice proceeds from a well meaning heart.

A good housemother wishing to perform her duties well must be healthy in mind and body and of an equable and reliable character. A vacillating person is not suited to the position of housewife: she must chose a simple, nourishing diet and accustom herself to bracing so that she does not become a victim to debilitation. Alcoholic drinks are to be avoided as much as possible.

Coffee and tea are both stimulants and often produce sad results.

The housemother should be especially prudent while in an "interesting condition" both as regards diet and bracing: in working she must not overexert herself for this is harmful.

I advise a woman in "the family-way" to take three or four half-baths in the week lasting only one or two seconds. I recommend also the knee douche and upper douche, but not more than one such daily, and to continue them until her confinement.

Such women remain healthy and strong and may look forward to their confinement cheerfully and gladly.

I have had many proofs that women who formerly regarded their accouchement with terror have had the best results from the water-cure.

Indeed I know of nothing which housewives and Mothers should lay more to heart than the necessity of bracing themselves, for upon this will depend the health and happiness of their posterity as well as their own.

A few days after the confinement the bracing may be continued in the same way.

I will furnish an example.

A Mother was assured by the Doctors that the next time she was in "the family-way" she would succumb. She lamented her fate to me, and I gave her the assurance that she need have no fear if she followed my advice which was to take four or five half-baths a week. This she did and her confinement passed off happily.

Another woman was quite inconsolable because she had already given birth to seven dead children. I gave her the same advice as in the former case and in six years she bore five healthy children.

Many Mothers will ask what they are to take for breakfast in the place of coffee. I answer: prepare yourselves for Monday's breakfast a good Brenn-soup*) which will nourish you and give you good blood.

On Tuesday a cooked bread**) soup also very nourishing: on Wednesday make a farinaceous soup of wheat, maize or any kind of grain which contains much nourishing food for the system. On Thursday another grain soup, but this time of rye, which is equal in nourishment to any of the former; and if on Friday you make a bread soup in water with some butter and onion you have again an excellent breakfast. On Saturday repeat any one of the above named soups.

As a change for Sunday make some strong malt coffee, and should a couple of coffee beans find their way into it, it need not be dangerous.

---

*) See page 75.
**) See page 75.

In certain conditions women should take a little lunch between breakfast and the mid-day meal: this should consist of a spoonful of milk and a piece of black bread: on no account take beer or Salami sausage.

Country people might take a little pot cheese with the milk. In the afternoon they should take a piece of black bread broken up in milk or they may eat an apple.

If the housewife has eaten a good breakfast and lunch she should choose something very simple for her mid-day meal: spices, strong acids, etc. must at all hazards be avoided.

Whether the meal should consist of meat or of farinaceous food is an open question; some people thrive on good meat diet, yet the Swabian women thrive equally on farinaceous ·food. I would rather recommend farinaceous food or even alternately farinaceous and meat diet.

Naturally the farinaceous food must be made of whole-meal and not of artificially prepared stuff, if it is to be nourishing and strengthening.

The simpler the diet the better; the more natural the more nourishing; the more artificial the worse for the eater.

The Swabian country women go at eleven o'clock into the kitchen and though they may have to cook for eight or ten people all is ready at 12 o'clock.

In Swabia the breeding of cattle has made great progress and one might think that this would increase the eating of meat among the people, but this is not so.

The Swabians have for generations past loved a simple, strengthening, farinaceous food.

If only I could succed in causing all housewives to practise a reasonable and continued bracing of the system; if I could teach them the superiority of simple nourishing diet and the avoidance of alcoholic beverages!

If I could induce Mothers to take my book on "The Care of Children" in their hands and to bring up their children according to its teaching, then I should carry with me to my grave the thought that I had been of use to mankind.

Second Part.

The Diet.

Good and well tested Cooking Recipes

for the

Healthy and the Sick.

Introduction.

~~~~

The first question we are called upon to answer is "What shall we eat and drink?"

Many people ask this who are anxious to choose what is best for their health and who yet in the end make mistakes.

In the present day it is really difficult to choose aright, so large is the variety offered to us. We will institute a trial of the things offered and see if even half will prove of service.

Inexpressibly manifold is our Organism, and the manifestations of life within it have a common principle as the source or fountain of life, namely the **Exchange of Material**.

If this exchange of material is turned aside from its course the natural result is illness, while its total cessation means death.

As we introduce nourishment into the body a certain process begins by means of which those stuffs which our ˙systemsneed are assimilated and worked up while the non-needful is thrown out as excrement.

How wonderfully our systems seek and desire that which they require, we see every day.

Quite instinctively the unreasoning creatures seek what is right for them: if they do not feel well they seek herbs which will heal them. **Instinct is their best Doctor!**

Thus hath God marvellously provided! so simply and yet so beautifully!

So-called culture has robbed us of the simplicity of our manner of taking nourishment which cannot be sufficiently deplored.

Nourishment should be exactly calculated to administer to our bodies much usable material; yet how much of this is lost by the artificial method of preparing our food.

Many people believe that there is much nourishment in the fine, dazzlingly white refined flour. That is not my belief; I see in it only flour dust and put no faith in it.

Formerly when refined flour was unknown one could get proper flour which was not impregnated with salts as now, nor as snow white.

One question I must allow myself to put which arises from my own constant observation. When children and even grown up people come from cities and towns into the country, the first thing they do is to ask for peasant's bread and sweet or curdled milk. Why is this, when at home they can have fine, beautifully baked, baker's bread?

Yet how good the children find this country bread; it is a pleasure to see them eat it.

The black bread is more tasty than the town bread. In this last I find the clearest proof that white and beautiful though it be, it does not satisfy as our black bread does simply because it lacks nourishment. The bread is better tasting and more nourishing when made of natural flour which is not scrupulously winnowed.

The fine extract is of no value.

I am sorry to say that our country people are no longer content with their black bread but try to get white, tasteless, non-nourishing bread. Just as coffee has gained an entrance into the poorest, houses so has refined flour.

No wonder that people are becoming weak and nervous when their nourishment is so adulterated!

The lack of usable material in the food we take causes great proverty of blood which is the cause of so many troubles.

Oh, that we knew nothing of bean coffee! But this miserable beverage has made its way into the houses of rich and poor, working mischief everywhere.

People will not believe that it is harmful notwithstandig that it contains a strong poison (coffein). It is palatable, therefore good. That it does harm to health seems a trouble to no one.

Let any one make up his mind to drink no bean coffee for a year and then to return to it after this lapse of time — he will then perceive the irritant effects of it.

I am firmly convinced of its harmful qualities and I do not stand alone in my conviction seeing that the best doctors reject it as a means of nourishment.

It may in certain cases be given as a medicine but even this I consider unnecessary. Men seek for delicacies which prove their enemies, and yet if one attempts to give a word of warning one is laughed at.

It is a comfort to feel that enough has been said and written against the pernicious custom of drinking coffee.

Equally injurious to our system is the use of alcohol, whether in the form of beer or spirits. There would be some excuse for those who drink them if they contained any nourishment, but little or none exists in either one or the other. People drink because it is the custom

and not merely to quench thirst, and the habit grows
and grows till people find it difficult to live without
stimulant. I am not opposed to beer drinking in moderation; it is the excess taken carelessly into the stomach
which does so much harm and causes the heart to throw
out too much water. Our carelessness gives the Doctors
a great deal of work.

To give children Schnaps, Beer, Liqueurs, etc. is I
consider an inexcusable attack on human life. Alcohol
in any form is for them simple poison and I call parents
who allow people to give it to their children neither
more nor less than assassins.

Coffee, beer, wine, schnaps irritate the tender nerves
of children and make them prematurely weaklings.

Children without doubt need a great deal of nourishment if their little bodies are to grow, and their food
should be varied if all the wants are to be supplied;
for example, the bones need chalk, the blood and muscles require iron and for the building up of the entire
organism nitrogen and carbon are wanted. Alcohol gives
children nothing that is of value to them. Again sweets
and pastry should not be known to children; they simply
spoil their stomachs and ruin their teeth; delicacies dear
and useless!

Dear Parents! Give the children simple, wholesome
food!

First Section.

Food for the Healthy.

I. Food.

It is impossible to supply the demand of the body for nitrogen and carbon by meat diet only. It would be necessary to eat a large amount of lean meat, as much as four pounds a day, in order to gain as much carbon as one needs.

Apart from the fact that one could scarcely take such a mass of meat for any length of time without disgust, the harmful results of the salts and extractive matter on the nerves and kidneys would make themselves felt.

It is needful that the nourishment should be composed of various nutritive elements. I do not like one-sidedness in food and therefore I do not agree with Vegetarians who constantly supply their bodies with too much of the same sort of nutritive matter.

Our bowels differ in form and size from those of plant-eaters, and we think, not without reason, that a great variety of nourishment is of use to us.

Too much and too constant a uniformity in food of a fluid nature and an entire lack of spice spoils the appetite, produces a distaste of eating and will even provoke vomiting.

One and the same food taken day after day has a bad effect on the stomach. A very great variety in diet is recommended to those who do not work hard and who are not much in the open air.

Merchants or hand-workers who lead a sedentary life cannot thrive either on the coarse food of country people or on a purely vegetarian diet, but the number of nutritious articles on which they may thrive is very great and depends upon individual needs.

The principal meals with us in Germany come at mid-day and in the evening. During the night the body does not use up so much material so that a portion of the evening meal is stored up in reserve and consequently the need for food in the morning is smaller.

Every healthy man or woman will know when and what he should eat and drink; the requirements of the system make this clear.

Meat itself, whether of one kind or another, should not be our only nourishment as it does not afford the system the requisite amount of carbon, therefore we must add such nutriment as will supply this lack. This we find in vegetables and various kinds of grain. Out of the last many good dishes may be made which are full of nutritive qualities.

Corn or wheat crushed and dried on a hot plate to a certain amount of crispness makes excellent grain soup. I prefer it to groats or oatmeal. These soups can be made so as to be very tasty and are much nicer than they look.

In grain there is a lot of albumen and carbonated hydrogen which make it very nutritious.

If one were to live on eggs alone it would be necessary to make one's will, for unless we could take about forty a day we could not get the proper amount of carbon and no one living could consume such a number.

This therefore is a new proof that variety in nourishment is absolutely needful.

Of the preparation of oats which are recommended and sold in shops I prefer above all oatmeal not too floury and rather coarsely ground. For little children one may use finer preparations of oats, but for grownup people the coarser kinds such as groats are best.

There is plenty of scope for invention in making dishes of oatmeal; very good and tasty puddings can be made of it, and if puddings of a finer quality are needed they may be made of whole meal and milk though of course they will not be so white as if they were made of refined flour. By using the oatmeal and whole meal we get the entire nutritive properties of the grain.

The importance of bread I need not point out as everybody recognises and prizes it as nourishment for rich and poor.

Bread is certainly the most indispensable nutritive article and it is the most ancient.

There are many kinds of bread and as many opinions concerning it. I always recommend bran bread notwithstanding the statement that even whole meal bread irritates the bowels. I agree that bread made of very coarsely ground husks may irritate but certainly bread made of properly ground whole meal flour will not have that effect. It has never to my knowledge caused diarrhoea; on the contrary, in costiveness of a determined nature I have brought about a daily regular stool by the use of whole meal bread. I maintain that the eater of whole meal bread is rarely if ever troubled with costiveness or piles.

Good whole meal bread prepared from wheat has
no acid and is therefore more bearable for a continuance
than the sourer rye-bran-bread.

In my recipes for whole meal bread I have stated
that yeast should not be used; in spite of this however people do use it, but it is wrong.

If whole meal bread is used in the making of soup
my advice is to cut up the bread small and fry it with
butter; or it may be toasted and in either case one
obtains a good tasting and very nourishing soup.

I also recommend that the bread be grated and dried
in the oven as a basis for soup. To improve the taste
a little spice may be added but very sparingly.

As nutritive food I recommend strongly green vegetables; they are of many kinds, rare ones and common
ones. Just as the dear God has taken care in all things
that the poor should not be stinted, so in this thing
also He has shown His fatherly love. It is true that
the poor or those with straitened means cannot imitate
the rich in buying truffles and the more expensive vegetables, but they can have at a small price an abundance
of nourishing vegetables such as potatoes, cabbages,
turnips and spinach. Of these I prefer cabbages because
of their good nutritive qualities; naturally they are for
healthy people; one would not think of giving them to
sick people or to those suffering from disorders of the
stomach with whom they would not agree any more than
the acid of herbs.

White cabbage can be prepared in many ways for
table and when variety is required one always turns to
this vegetable.

A Doctor once said to me "If there were no cabbages there would be many more invalids" therefore
cabbage must have a good effect on the system.

On the whole all plants which we use as vegetables
contain materials necessary for our system and which it
at once absorbs.

Why do we recommend green vegetables, especially spinach to anaemic, and chlorotic people? Simply because they are ferruginous (possess iron) and therefore excellent for the formation and increase of blood.

It is evident therefore that our food should be varied and at the same time simple.

In the present day it is quite easy to obtain all kinds of vegetables, and they are familiar in almost every household, but none more so than cabbage.

If I speak strongly in favour of cabbage it is because I know of no vegetable from which one can make so many dishes and suit so many tastes.

II. Vinegar, Salt, Spices.

I must now speak of something very important, viz. the use of spices, in which many mistakes are constantly made.

Cooking-salt which is used in the preparation of food renders the latter tasty, improves the digestion and has a preserving effect.

Salt is found in a lesser degree in drinking water, in a larger degree in nourishment itself, as for example in meat, eggs and milk, and in many other articles of food, yet it is often added to food in large quantities: and I maintain that the oversalting of food is not beneficial to the body. As a proof look at people who live on animal food. Some people have an aversion to salt, and I know many who do not take it at all and yet enjoy good health. There are even Vegetarians who say that with their diet little or no salt is required. I know cases in which people who were not in the habit of eating salted food when forced to do so, got blisters on their lips which lasted several days and were quite painful.

3*

People are not wise who take salt with their food in **large quantities**. It produces inflammation of the stomach and bowels, vomiting, diarrhœa and even death. Salt has a strong effect on the kidneys and skin (perspiration); it produces atrophy, scurvy, skin-disease with itching and catarrh of the mucous membrane.

It is therefore well worth the trouble of warring against a superfluous use of salt considering all I have said.

I am bound to speak out the warning and I trust my words will not be spoken to empty air.

Another kind of salt used for the curing of meat is saltpetre. I do not care for ham nor do I prize it as highly as many do, probably because I know that a large quantity of salt-petre is used in the curing of it. I do not think I am wrong when I assert that salt-petre is injurious to health. It might be permissible in very small quantities, but considering that ten to twenty grains are a strong poison I think we must come to the conclusion to avoid salt-petre.

Even taken in homeopathic doses it can have no good effect on the system.

Now-a-days one prepares so many artificial nourishing articles of food without pausing to consider whether they are suitable for our bodies.

The adulterated preparations of nutriment require so many chemical adjuncts that one is bound to experience a certain mistrust of them. I attribute the many diseases of the stomach and bowels to adulterated food and to the excessive use of beer, etc.

Again **Vinegar** is employed in food and readers may think I shall deal less harshly with this article, but they are mistaken.

Vinegar is considered indispensable in the kitchen; whether it has a right to this distinction or not is an open question.

If the vinegar is pure and unadulterated then I allow the reasonable use of it; I, however, emphasize the **reasonable** use. Why do I say this?

Because vinegar taken in too large quantities is harmful. A proof of this is patent in that the excessive use of vinegar long continued causes the most beautiful complexion to fade.

Unfortunately there are stupid people who, in order to acquire the pale complexion considered now-a-days aristocratic and elegant, plunge themselves into the danger of losing their life or at any rate shortening it.

Used in moderate proportions vinegar has a preserving and conserving effect; it also makes the food softer and pleasanter to the palate and removes thirst. The best is certainly fruit vinegar and I consider it the most wholesome.

If one adds aromatic herbs or fruit to vinegar and allows it to stand for twenty-four hours, the result is a good herbal vinegar which gives a pleasant taste to food.

Used in moderation and with discretion vinegar does not hurt; it has at any rate the right to a place in the kitchen.

I have spoken of vinegar as for use in the kitchen, it still requires to be noticed as a **healing agent**. Vinegar also is used with water in ablutions, cold compresses, hot bandages, etc.

Vinegar has the power of preventing putrefaction and is used as a gargle or paint in diphtheria.

I am very partial to vinegar as a means of healing, and certainly it deserves this recognition: still be careful always to get good unadulterated vinegar! In my book "Thus shalt thou live" I have given the recipe for preparing vinegar from fruit. I will now close this section on Vinegar with the warning "Be sparing in the use of vinegar in food and do not use it in large quantities as a beverage even when mixed with water!"

As I am in the way of warning I will add yet a word on spices. One uses spice in food to make it tasty and piquant, and I have nothing to say against it so long as it is used in moderation. Spices play a large part in the seasoning of our food.

Too strongly spiced food produces distaste which may increase to sickness and loss of appetite. It is quite a mistake to consider spices harmless; by far the largest number of abdominal diseases, especially in people of a better class, arise from the overspicing of food.

The doctors discover this daily. Liberality in the use of spices, salt and vinegar introduces many diseases into our homes which might have been easily avoided. Nor is the evil confined to abdominal diseases but includes enlargement of the liver, costiveness, piles, swelling of the bowels with catarrh in apparently healthy people who are simply overspiced and nervous.

She is not a good housewife who sends to table sharp and piquantly spiced food, on the contrary she only deserves the name who produces tasty and yet simple food with very little spice.

Such is my opinion as to spices; nor do I stand alone in my conviction, for many very good Doctors think with me.

I warn my patients at all times against the misuse of spices, salts and vinegar, for I know for certain that to have none of these is healthier than to have too much.

Work, moderation and regularity in eating, drinking and exercise in the fresh air are the best spices.

Up to this point the remarks I have made on nutriment have been somewhat compressed. I have only touched on such points as I considered needful for you to know in order to live.

III. Fruit.

Fruit being a very important article of food requires a more minute description. It is indisputably the most agreeable and the most ancient of our articles of food. Going back to the earliest days of the world's history we learn that Adam and Eve lived principally on fruit and from that time to this it has been an important factor in the daily food of man. It is pleasant to the taste, it has a delicate perfume and possesses a combination of acid and sugar all of which make it most acceptable.

Good Mother Nature has taken care to provide us with fruit during the greater part of the year; each season taking its part in the production.

Strawberries come early before the great heat sets in and one is not so thirsty; this kind of berry does not contain so much acid; then follow raspberries, bilberries, cherries, currants and gooseberries. At this time the heat is greater and our blood is thicker, and so the agreeable thirst-satisfying acids of these fruits are exactly suited to the time of year.

Later other kinds of fruit follow: apples, pears, plums, apricots and peaches, nor must we forget blackberries.

They say of fruit that it produces illness but I am not of this opinion. I do not deny that the misuse of fruit produces disease, but are we bound to misuse it? God did not create fruit for that purpose, He has created nothing for misuse.

I am of opinion that the right use of fruit reduces illness rather than instigates it.

Decidedly it is better to eat good ripe fruit than to swallow a variety of chemicals.

The consumption of unripe fruit is both objectionable and dangerous because its acids are not wholesome.

Fruit has great nutritive value and in its fresh condition is not behind our best vegetables. Of course one must not look on it as a sole means of nourishment; no sane person would think of living on fruit alone. Nor must one regard it merely as a luxury because it is really nutritious.

Liebig, a very celebrated man thought most highly of fruit, and not without reason. How agreeable to the taste is a piece of wholemeal bread eaten with an apple in the course of the morning or the afternoon. I recommend this last to the anæmic.

To people who indulge in drink I strongly recommend fruit diet as a sure method of healing.

With fruit many a charming dish may be served up; I wish to say a word about preserved and dried fruits.

Preserved fruits are known to be very cooling and therefore suitable for invalids; still one must discriminate when and in what cases it is permissible to give them.

Dried fruit has even a greater nutritive value than fresh fruit and is thought to be easier of digestion. Dry the fruit in the sun where possible, it is equal to a second ripening.

Wine is made of fruit and receives the name of that employed in its preparation as a rule. That which we make from apples in England is called cider though in Germany it is called apple wine. This has an aperient effect and therefore is recommended for costiveness. Very fat people drink it in order to get thin, but I advise them to be careful not to take it in too great quantities at first as it sometimes produces colic and diarrhœa.

I do not think it necessary to treat of the various kinds of fruit as kernel fruit, stone fruit, berries and skin fruit, as there are many books full of information on the subject. My sole desire has been to bring forward the use and value of fruit in order that all who read this book may realize its double value, first as a means of nourishment and secondly as a curative agent.

I believe that if more fruit were used and less spice the habit of taking alcohol would considerably decrease.

IV. Fine Pastry and Confectionery.

I cannot deny myself a few words on so-called fine pastry and confectionery.

Art makes so · many things in this way that one doubts the necessity of such dainties. They are after all only for Dessert as tit-bits, and not as nourishment. But they are partaken of greedily simply because they are supposed to belong to "good form". The stomach anyhow is already overfilled, but that is no matter because the succeeding coffee and a small glass of brandy equalizes everything and assists digestion as people usually say. Many a girl lets the real meal pass and saves her appetite to attack the dessert and swallow the sweets wholesale.

The value of such pretty confectionery is naturally very great in the eyes of lovers of such things. Children who love comfits and such like sweets have abundant opportunity of procuring them and so spoil the stomach and ruin the appetite for strengthening food: If one will use these valueless dainties as dessert, they should take care not to let them take the place of the meal itself.

Poor people who cannot afford these dainties are in my opinion much better off, for a piece of their black bread has more value than heaps of such elegant confectionery.

A mother who, out of pure tenderness, gives her children sweetmeats and sugar plums I consider unworthy her position and she should on no account think she is rendering her children a service. On the contrary she is weakening them and later on she will reap the fruit of her indulgence.

One may now and then give children a little treat so as to make them more diligent and attentive, but it must not happen too often, otherwise it becomes a custom against which it will be difficult to fight.

V. Beverages in General.

Alcoholic specially.

Water, as given us by God, is the best beverage and I think that a tolerably long time passed before the human mind conceived the various sorts of drinks now in fashion.

Wine is an exception for we know that Noah prepared wine and was the first to prove its potency.

Before touching upon the different sorts of beverages I will give my opinion of beverages in particular.

My principle is, drink when you are thirsty and not too much at once.

Never drink while eating; rather drink before or after doing so!

These are old rules of health, but like many others now-a-days they are no longer observed. "Much drinking is a custom and no natural disposition" as it is wittily expressed!

The custom of consuming a quantity of beer has slipped in on account of modern social life and people pride themselves on being able to swallow the largest possible amount.

When one reads the enormous number of gallons of beer consumed in a single town and divides them among

the inhabitants in order tho see the quantity each consumes, we are startled into asking "is this possible"?
and if so is it right or reasonable? Certainly not.

You have only to look at one of these topers to
see the effect the drinking has had upon him; his complexion is red, blue and congested. It could not be
anything else; for the heart must pump out the superfluity of fluid from the blood and dispose of it somewhere; the kidneys, too, get more work to do and are
strained beyond endurance. If one considers that the
human organism may be compared to the highest kind
of skilled machinery in which every disturbance, no
matter how small, causes discord, one will at once understand that by the excessive and continued use of alcoholic drinks the whole system is put into disorder, a
mischief by no means to be underrated. The excessive
drinker brings upon himself many diseases and infirmities.

It seems scarcely possible that a creature gifted
with reason as man is, could commit such a mistake
seeing that in other ways he takes such precautions for
his health. One preaches, however, to deaf ears when
one touches on the subject of drink, notwithstanding
that every day's experience shows us the consequences
of this want of moderation.

Wine.

I admit the truth of the proverb that "Wine rejoi-
ces the heart of man" although I drink no wine myself
or at most only sip it on very rare occasions.

Wine is undoubtedly one of God's gifts, and a natural beverage as long as it remains pure and unadulterated. Although it is preeminently an article of nutriment yet it is largely made use of for medicinal purposes.
Enjoyed in moderation it assists digestion and affords
stimulant in exhausting illnesses giving strength to the
patient and helping towards his cure.

The misuse of Wine, however, does away entirely with its good effect.

I do not approve of so-called "Made up Wines", for they are harmful, neither can I praise sparkling wine.

When on very rare occasions I prescribe wine it is always a simple light sort, and then I only give it to old or very weak people; **to children never,** for to them, especially to those who are nervous and excitable, it is poison.

I know that it is the fashion at the present time to give children beer and wine; the consequence is that nervousness increases and the weakness of the next generation will be more marked even than that of the present.

Beer.

The reproaches constantly thrown at me for so strongly forbidding the drinking of beer do not make me at all nervous; I can afford to bear them lightly; for he laughs best who laughs last. Beer has many nutritious qualities ascribed to it, but with what right?

Beer is an irritant for it contains alcohol and that in the lightest and thinnest of beer. It has been told me that people must have beer when at hard work in order to keep up their strength. This I do not believe. When I look at the Italian workmen and observe how very hard they work without drinking beer, I am confirmed in my opinion that it is unnecessary. Often a whole week passes by without their touching either beer or wine for they are very poor, but they eat heartily, and it is the food that gives them strength, not alcohol which quickly evaporates, nor beer.

These are but a momentary stimulant, a blazing fire which dies out again as soon as the alcohol evaporates. If one would spend on bread that which is spent on beer, one would at least have nourishment; as it is the money goes to buy a quickly passing stimulant.

1 am quite amazed that there should be people who pride themselves on having drunk such and such a quantity. I pity them; for they pride themselves on something of which they should really be ashamed.

What a senseless custom is that of students who even institute special trials of strength drinking one pint after another as if that were any honour. It never seems to strike them that they are thus injuring their health.

Let those who really think they cannot live without beer at least be moderate and remember that all immoderation brings punishment in its wake. Drink no more than is absolutely needful to quench thirst and a very few half pints will do this.

Drinking much is a habit and grows upon one, the more one drinks the thirstier one becomes, for drink excites thirst.

Brandy.

The continued misuse of alcohol has visibly heavy consequences brought about by sympathy of the whole system.

The central nerve system, the heart, kidneys, liver, arteries, stomach and intestinal canal must all suffer from this vice.

If only one of these important organs is attacked there is danger to the whole system.

Small quantities of alcohol may, as one says, bring greater heat to the body, but still I can never agree to its consumption nor will I recommend it.

Unfortunately the misuse of alcohol is at the present time very great. "Well" says some one, "what can a poor man do if he has but little money? He must buy himself a dram of brandy in order to get a little warmth into him, and so be able to work."

This speech is not satisfactory. A small glass of schnaps costs sixpence ; naturally one drinks two so as to get more heat and that makes a shilling.

Surely the Workman could with this money have bought something at once tasty and strengthening for his work!

Certainly he could.

People who worship brandy take very little nourishing food, a circumstance which is dangerous when in combination with the misuse of brandy especially of the commoner kind.

Regarded from a dietetic point of view brandies are of small value: they may be agreeable, but they are very dear articles of consumption.

The many liqueurs, health essences and such like productions are only secret grave diggers bringing the consumers to an early grave.

If then the drinking of brandy is so dangerous to grown people, how much worse must it be for young people and children!

What an immense responsibility parents voluntarily assume when they give their children spirits to drink!

I do not stand alone when I warn you against the use of brandy specially and the misuse of alcohol generally: I have the support of all good Doctors and of thoughtful reasonable people; even the Church utters its warning in the same direction.

Why are we all of one opinion on this matter? Because its consequences lie so plainly at hand and so many families suffer from them.

How much domestic happiness, how much health is ruined by the misuse of alcohol!

Does not this mean an immense responsibility?

Sapienti sat!

Coffee, Tea, Chocolate.

All of these are articles of consumption and come under the term beverages.

I have already written and said so much on these things that I will not linger over them now. I recommend **neither tea, nor coffee, nor chocolate.** Quite lately a patient reproached me for being so dead set against coffee, "it could not," he said "be as dangerous as all that". I however stick to my opinion and with reason. No one will deny that the misuse of it by women a venges itself heavily on the nervous system and accounts for the number of weak nervous children in the present day.

A doctor says "never give children coffee during the first four or five years". He would not give such a direction without reason. Chromic yellow is poisonous and yet it is frequently used to colour coffee.

It is the opinion of many that the good qualities of coffee out-balance the bad.

I will allow that with many people the moderate consumption of coffee does not act badly; still in the majority of cases its action on the nerves is too strong. I therefore warn children and also women in an "interesting condition" against drinking coffee; its effect is irritating.

Tea also belongs to the stimulating articles of consumption. It is very much in fashion as an article of diet, but even those who approve of it admit its exciting effect on the nerves and therefore warn you against its misuse which is a proof, if one were wanting, that I am right.

My long experience and many observations have convinced me that these beverages are harmful especially when their use is abused.

Chocolate is far less irritating than tea and coffee, therefore I do not oppose its use so much.

In Southern France cocoa is taken as a remedy for diarrhoea. Acorn-cocoa taken with cinnamon is also good for the same.

On account of the amount of saccharine matter which chocolate possesses it cannot be taken for any length of time without producing acidity in the stomach.

Second Section.

Food for the Sick.

~

General.

Whereas with healthy people the need for taking nourishment depends on the conditions of work, rest and climate, of age, sex or habit, with sick people it depends upon the power of digesting or assimilating the nourishment.

The kind of decomposition of the tissues and the degree of capability in the digestive organs act differently in sick people. It is therefore necessary first of all to determine how far the digestive organs are disturbed in their functions.

A patient suffering from fever must certainly be treated more cautiously than one who has no fever. According to the illness so must the diet be regulated.

Just as diseases are manifold so also are the nutritive materials which should be given to the invalids.

In cases of feverish, acute illnesses the food should be as light as possible because the digestive powers are more or less severely affected.

Sick diet should never be thick and compact but on the contrary of a fluid nature.

Again food should never be given too hot to fever
patients, it should be rather of a lower temperature, and
beyond this there should be a variety of food suitable
to the class of illness and the capabilities of the digestive organs.

I will also take this opportunity of calling your
attention to the fact that a certain order should be observed in giving food to invalids.

Attendants and Nurses often worry the patient to
eat something thinking that by forcing him to eat they
are doing their duty, forgetting that the digestion requires
extreme regularity in eating and drinking, a course to
be strongly recommended.

It is better for any one who is seriously ill to eat
little and often rather than much at one time.

With some invalids it is necessary to force nourishment upon them; this is specially the case with the
delirious.

In the case of health-restoring sleep one should not
wake the patient for the sake of giving nourishment,
but rather wait until he wakes of his own accord.

Again I earnestly press upon you that nourishment
should be given during the night to persons who are
seriously ill; the interval from seven or eight o'clock at
night till seven in the morning being much too long.

It is very injurious to give solid food in feverish
cases and especially to patients suffering from Typhus,
Diarrhoea and Inflammation of various parts of the
system.

Solid food is useless because the patient cannot
digest it without the necessary gastric juice which, in
the diseases mentioned, is lacking.

Again the nourishment of fever patients must be
regulated by the age, the strength or the weakness of
the constitution; for instance an old person has from the

outset less strength and power of resistance against feverish symptoms; it is just the same with anæmic, consumptive, weakly people and drunkards, all of whom have enfeebled systems.

The nourishment to be given in all cases must be regulated by the disease, nothing must be given that the patient cannot digest.

In order that there may be no mistake I will set down directions as to food in the various illnesses.

Peritonitis.

In acute peritonitis give filtered barley water or gruel, groat-water-soup in small portions frequently.

In case of constant vomiting one must stop giving nourishment for a time.

The beverages should be cool.

When the fever first ceases give cornflour soup mixed with milk or the yolk of an egg.

When the sickness and swelling of the intestine canal no longer exist you may give potato broth with milk, rice milk and well chopped or minced meat.

Gradually you may give food a little more solid but not until all fever and irritability of the abdomen have disappeared.

Abdominal Typhus.

Just as in Peritonitis one has to observe caution in the diet so also in abdominal Typhus only with this difference that one must keep in view other important circumstances.

In Typhus the fever lasts much longer and therefore exacts more from the patient's powers.

4*

It must not be forgotten that in Typhus the fever is but slight in the beginning and gradually mounts higher and higher. One circumstance in this fever must never be disregarded, viz. that the ulcers in the bowels which nearly always appear in severe typhus are not healed when the fever ceases: therefore it is necessary to avoid everything that increases the inflammation of the mucous membrane and might prevent the healing of the ulcers; on the other hand, however, it is necessary to give the patient as much nourishment as he can bear. Let a typhus patient drink a few tablespoons full of water every half hour; also a little sugar water or bread water, (that is drinking water in which bread has been soaked), is beneficial with an addition of a little brandy or red wine.

Fruit syrups are only allowed in small quantities when no diarrhœa exists.

Let the first food given be soup with egg, thick soup, malt coffee rather strong so that the patient may imbibe a good deal of the extract of malt.

Solid food is strictly prohibited: even in the convalescent period one must be as careful as possible on account of the ulcers in the bowels which are still in the process of healing.

When the fever has completely disappeared you may begin to give cornflour-soup, milk soups, potato soups, bread soups and finally minced meat easy of digestion.

I would remark here that it is wiser not to yield to the feeling of hunger often felt by the convalescent patient because as a rule it is false hunger.

Not seldom he swallows a great deal at one time, and the result is a relapse which is a sign of returning fever.

Dysentery.

In Dysentery nourishment must be regulated not only with a view to the digestive capabilities but also to the local suffering. One has to remember, as in Typhus, that with the cessation of fever and improvement in the digestive organs a complete healing of the mucous membrane does not immediately follow.

Dysentery is difficult to manage even when it is but a slight attack.

On account of the great irritability and unusual sensibility of the intestine canal do not give the patient ice cold water and not much at one time because it causes pain and cramp. Water which has stood for a short time is best.

Bread water, almond milk, rice water and weak Tormentilla tea are all good for the patient if they are taken in small quantities.

I recommend as nourishment barely water, gruel, rice-flour soup and groat-soup cooked with milk.

Do not give the patient milk pure from the cow the tendency of which is to cause Diarrhœa. It is only by degrees when diarrhœa ceases that the patient may begin to take more nourishing diet.

During the illness no effervescing water should be drunk, and acid beverages and spices must be strictly forbidden and except the patient has a very weak action of the heart he should not take alcohol.

People are much too prompt with the brandy bottle. I do not consider this right, for it excites the patient and often causes him more worry than rest.

Should vomiting appear with the dysentery, one of two things must have happened: either soil has accumulated in the intestine canal or some mistake has occurred in the diet. In either case give the patient ice water to drink and for nourishment on y thick soup.

When however the diarrhœa and vomiting cease more nourishment may be given.

~~~~~~~~~~

Acute Gastric Catarrh.

In this disease it is not difficult to bring about a cure in a short time. The invalid should observe a strict diet and take no solid food; he may have meat soup with egg and keep to that only until the digestive powers improve and appetite returns. With such treatment the chill passes in three or four days. People who suffer from acute gastric catarrh generally have a sensitive stomach.

~~~~~~~~~~

Chronic Gastric Catarrh.

This is a very unwelcome guest, not only on account of the suffering it entails but also because it is exceedingly wearisome and difficult to cure.

Also in chronic gastric catarrh it is not easy to prescribe the proper diet; indeed the patient himself, if he will give his attention to it, can best say what his stomach can stand.

Leguminous articles, such as beans, peas and the like, should be avoided because they develop gas and irritate the stomach both of which are injurious to the sufferer. The Doctor settles a regulation diet which for poor people is simply impossible. This and that is forbidden and a variety of things recommended. Rich people of course can manage to procure these but poor people, if all these were absolutely necessary, would suffer cruelly and never get well. **But it is not so.**

The poor as well as those who are well off may without doubt recover health by means of the simplest nourishment.

I do not myself value the regulation diet which is chemically analysed and weighed nor do I believe that to satisfy the stomach it is necessary to buy expensive things.

Each individual requires a separate treatment, for what will be good for one person may disagree with the other. Many patients are of the opinion that if there is no appetite it should be stimulated by piquant food. This is however a great mistake, for spices irritate and inflame the gastric mucous membrane and so retard the cure.

The simpler the nourishment the easier the cure.

Those who desire to heal chronic Gastric Catarrh should avoid loading the stomach with an excess of food and drink and by too frequent meals. Nor should strongly spiced food be taken, nor a one-sided diet.

Greasy food readily becomes rancid in the stomach and acid food easily ferments thereby causing flatulence.

A tablespoon full of pot cheese taken every two hours assists greatly in improving the condition of the stomach.

Milk diet, light farinaceous food and carefully prepared animal food may be recommended to gastric invalids. Sweetmeats and confectionery do not suit a sick stomach, therefore away with them!

Many complain of gastric troubles who are often themselves to blame for them and if they find no cure you will discover that they are in the habit of tickling their palates and filling their stomachs with all sorts of things whether suitable or not.

Gastric Inflammation.

Another trouble is that of gastric inflammation which is accompanied by fever. Although the diet given is almost identical with that recommended in gastric catarrh one has also to operate locally by cold compresses on the stomach and these must constantly be renewed.

Although the inflammation may in this way be rapidly reduced the patient must remain in bed.

When the chill and inflammation have passed the food supplied may gradually grow stronger and more solid, but the greatest care must be taken not to yield entirely to hunger and not to eat too much at once.

Costiveness.

Ordinary costiveness arises from various causes.

With one person it may be ascribed to faulty nourishment, with another to flaccidity of the muscular system, or contraction and variation in the position of the intestines.

People afflicted in this way generally take a strong aperient in order to overcome the evil; this, however, is a mistake against which I cannot too earnestly warn you.

The continued use of violently acting medicine renders the bowels more flaccid and more incapable of independent action so that aperients have to be taken again. In this way the cause of the suffering is not removed but rather made worse.

I recommend merely a tablespoonful of water every hour not for one day only but until the action of the bowels is regular.

The diet must be so arranged that all articles of food containing tannin are absent, for instance, red wine, biscuit, cakes or bread made with yeast, confectionery, bilberries and the like.

Honey, butter, milk, rye bread and tablespoonsful of red sour kraut are to be recommended. Medicines do not help much so we will leave those alone.

Chronic Diarrhœa.

Chronic diarrhœa is exactly the opposite of costiveness. One can scarcely imagine anything more disagreeable than an obstinate diarrhœa. It renders one powerless, tired and incapable of work.

To cure diarrhœa, in contrast to costiveness, it is necessary to reverse the diet, that is to say, what one takes when costive must be avoided in diarrhœa, viz. aperients such as coffee, honey, sugar, fruit, fruit-wine, effervescing waters, acids, milk, pickled food, etc.

For nourishment take acorn coffee, thick soups of barley or rice, farinaceous soups (but no farinaceous puddings) and bilberries.

Also the quantity of the nourishment taken must be moderate if a cure is, desired, therefore I would advise that the meal times be rather more frequent than that a great deal should be taken at once. At the same time the patient should take daily one or two cups of decoction of bark of oak and tormentilla root.

For children acorn coffee is greatly to be recommended as also are small quantities of chalk flour dissolved in water.

Care should be taken that children have their meals regularly for in many cases this will cause the diarrhœa to cease.

I do not advise you to give wine to children; it is possible to accomplish something with simpler remedies.

Hemorrhoids.

Let the sufferer from this disease direct his attention towards getting the bowels gently open so as to avoid costiveness.

The diet should never be too heavy or too dry but consist rather of simple household fare. There may be variety in the food if leguminous articles, much black bread, heavy wines and strong coffee are omitted.

Meals with numerous courses must be avoided.

I recommend walking and sawing wood as a daily exercise.

Kidney Diseases.

In kidney diseases the digestion is always more or less disturbed. Loss of appetite even to the extent of loathing food sets in and vomiting and diarrhœa are of frequent occurrence.

I recommend to those suffering from this disease an abundant milk diet but at the same time very little meat. Many Doctors forbid meat altogether, but with this I do not agree.

As patients afflicted with kidney mischief suffer much from loss of albumen, a diet must be chosen which will supply this in a moderate degree.

Strongly salted ham is harmful to the patient and should be avoided.

I do not forbid vegetable food, indeed the sick person may, without fear, take mashed potatoes, rice and fresh vegetables. Care however must be exercised that too much is not eaten at one time; I would rather that the nourishment should be taken frequently in small quantities, provided always that the meals are regularly given.

Alcoholic beverages and strong seasoning and spices must be strictly avoided.

It frequently happens that patients suffering from this disease vomit at every meal; in this case I recommend cold milk in small quantities and a tablespoonful of weak worm-wood tea every hour; also a tablespoonful of pot cheese every hour would be of service.

Glycosuria; a form of Diabetes.

For patients suffering from this form of Diabetes Doctors generally prescribe a meat diet only which is the exact opposite of that ordered for persons afflicted with kidney disease. How far they are right I never can quite determine.

I have given a variety of food in diabetes and yet the patients have recovered. It seems to me that to give only meat weakens the system, which should be the more avoided seeing that the loss of saccharine matter is great.

Beside the invalid could not endure living on meat only for very long, the stomach would become inflamed and gastric catarrh would result.

Sweet things such as puddings, confectionery and the like are very unsuitable nourishment in this disease and I as a rule condemn this class of food because it is of no service to the stomach.

Sugared coffee, milk and honey should all be avoided. Large quantities of bread do not act favorably.

Fruits containing sugar with the exception of currants, raspberries, blackberries and sour cherries must be equally avoided.

I consider the Doctors' prescriptions which only permit patients to have meat and forbid bread, farinaceous food, milk, etc., much too severe a doctrine. It is said "Science has discovered what is best to give in this illness". Well I admit this much, viz. that science has discovered much that offers a standpoint from which to attack disease effectually, but I cannot conceal from myself that it is precisely in this illness that so many of the Doctors contradict themselves or at least, are not clear as to whether this should be allowed or the other forbidden. Let that be as it may; I decidedly affirm that a diet rich in variety is the best and the results prove me to be right. That this would be the case without application of water I do not believe, for water must have its place therein as a curative agent.

For great thirst let the patient drink fresh water in small quantities. Tormentilla root decoction is also good as a beverage; thirty drops of tormentilla tincture taken thrice daily in a tablespoonful of water quenches thirst. It is quite a mistake to think that large quantities of water quench thirst. For the moment it certainly answers, but thirst soon sets in again and then one is obliged to pour into one's stomach the same quantity of water as before. The loading of the stomach with so much fluid is, however, on every score greatly to be condemned.

Small quantities taken frequently are much more effectual in quenching thirst than large.

I do not recommend wines. In the first place most of them contain a good deal of saccharine matter except of course those that are perfectly pure; secondly they are not so necessary but that the poor may spare themselves the expense.

It holds good here as elsewhere that the simpler one makes the road to health the surer will be the result, and the more complicated the means the more uncertain the success.

It would be extremely difficult, as a rule, for poor people to cure themselves of diseases, were it not that Mother Nature is so good and generous in supplying them with simpler remedies than those prescribed by the Doctors.

It is quite impossible in this form of Diabetes to follow an unvarying strictly defined diet, one must judge according to circumstances and experiences.

Bladder Troubles. Chronic Catarrh of the Bladder.

Sufferers from diseases of the bladder must be forbidden the consumption of wine, beer, cider, pickled and salt food, pungent spices, mustard and horse-radish.

The nourishment may be varied but it must not be prepared with seasoning or pungent spices, otherwise it would irritate the system.

Very salt ham must not be eaten. I recommend malt coffee as a beverage. Mineral waters are not suitable for bladder diseases. A milk cure would be decidedly the best for this trouble.

A person suffering from disease such as this would do well to avoid entirely all alcoholic beverages like beer, wine, cider, etc., even when he feels himself cured; by so doing he will spare himself much pain.

In weakness of the bladder and in bed-wetting the above advice holds good with this difference however that only dry food should be given at night until the evil is removed.

Poverty of Blood and Scrofula.

I have often put the question to myself "Why it is so many people suffer from poverty of blood and why one sees so many scrofulous children?" It might be truly said that half mankind suffers from this evil, and I cannot see what else can be expected when one takes into consideration the modern way of living.

The topsy turvy kind of life pursued by the present generation is answerable for it in a large degree and other things help, for instance, the haste with which one works, the struggle for competence, the crowded dwellings in big towns, the want of air and light in the houses, and scant as well as badly prepared food.

The passions of men are the sources of disease, causing weary sickness to whole generations.

Of course poverty of blood (as well as other troubles) arises sometimes from loss of blood by accidents, still these sort of occurrences form the minority while the causes already given represent the majority.

The question is how can we be of use to such invalids so as to bring back strength, improve their blood and give them again the days wherein they enjoyed life?

The chief thing necessary is to give them simple but strengthening food. The usual cry from the poor in answer to this is "We have not the means to procure good strengthening food, we are badly looked after".

This is not true, for in simplicity lies curative power not in riches. Unluckily the opinion has crept in that only the rich can be well, for he alone can procure everything that he wants to establish his health.

The complaint is not justified.

Even if the poor must go without some things yet contentment will compensate him and he must know that with the rich it is not all gold that glitters.

Millions of money will not help a man if his system refuses the healing process, or if the curative agencies shirk their duty.

It is not always a great happiness to be able to command large means, for then we frequently employ curative remedies which count for nothing and do harm rather than good.

It is possible to assist the system during the healing process quite as well with simple remedies as with expensive which last are in many respects useless.

The cure of poverty of blood and scrofula is a long process and the arrangement of a suitable diet is a very important part of it.

All sweetmeats and confectionery, unripe fruit and many alcoholic drinks must be strictly avoided.

The food should be varied and I recommend green vegetables, milk puddings, pot cheese, grated cheese, whole meal bread, strong broth, grain soups and animal food in small quantities.

Fruit is good for anaemic persons but it must be ripe.

For persons whose stomachs are weak I recommend only easily digested food such as pot cheese, milk and strengthening broth.

Care must be exercised in the diet that it not only strengthens the system but helps to remove the causes of the disease.

If a person suffering from poverty of blood and scrofula wishes to get well he must not live in a gloomy, unhealthy dwelling which lets in but little light or air; and he must give up everything which he knows will hurt his health, otherwise no remedy will be of the slightest use, he will remain a wandering corpse even with the best food.

I wish very briefly to point out that in sexual troubles which greatly injure the blood there is nothing better than a simple nourishing diet. Formerly Doctors gave a very strong diet without regard to the person's powers; now the diet is chosen according to the capability of the patient's digestive organs.

I have always been of opinion that the blood may be improved by proper nourishment regularly given, but I emphasise the fact that I am not and cannot be in

favor of alcoholic beverages or irritating food and
spices.

In such illnesses moreover radishes, onions and asparagus must not be eaten.

Gout.

It is a very general opinion that gout arises from
excess of eating or from injudicious nourishment. In
many cases this is correct but not in all.

I think that in gout different factors lend their
assistance in developing the disease.

Great drinkers suffer specially from gout and so do
gourmands whose meals consist of rich foods. On the
other hand, damp houses foster the disease.

It is impossible to prescribe a definite diet in gout;
a mixed one is certainly the best.

The toper must slowly reduce his quantity and the
lover of heavily laden tables must be content to sit down
to simpler food, while he who has a damp house must
look out for a dry one!

Vinegar and spices should be avoided.

Fruit is good for those suffering from gout if their
stomachs will bear it. Gorging and extra meals must
be strictly avoided and I warn them against too much
animal food and advise instead more of vegetable diet.

Feverish Attacks.

If the invalid have fever it is, as a rule, because
the digestive organs are disturbed and this results in
loss of appetite.

One would never think it possible that nurses would
give such invalids solid food or drink which to start
with are known to be injurious, yet impractical people
do the most incredible things. They force the fever

patients to eat and quite believe that the eating will bring about the cure more rapidly. This of course is very wrong.

When fever appears give the patient only fluid or pappy food which can easily be worked up by the stomach, and when the disease is really established arrange the nourishment accordingly.

It does not hurt the patient to observe a very strict diet for a few days; on the contrary it will often avert a worse turn to the illness.

Should the patient refuse to eat for a long time, he must be treated firmly and compelled to take at least a small portion of nourishment every two or three hours.

Patients who are unconscious or delirious must be supplied with nourishment by the attendants in small quantities and often, whenever they find the opportunity.

Beverages in any case must not be given in large quantity; rather very little and often which quenches thirst more surely.

For the rest, thirst is much rarer during the application of the water cure than in allopathic treatment.

Convalescent Diet.

The period at which to begin the convalescent diet is not always correctly chosen especially when we take into account that, for instance, in Typhus the gut is very inflamed long after the fever has ceased.

It is imperative to notice at this time how far the digestive organs are working and whether they are still very sensitive. As a rule after severe feverish illnesses the mucous membrane of the stomach and intestines is very weak and sensitive and necessitates the greatest caution in the giving of nourishment. We constantly find that invalids arriving at the convalescent stage complain of difficult action of the bowels. Why is this? Simply because the gut is still weak and languid and does not act properly.

Therefore one has to determine for each individual separately when the time comes for changing the patient's diet; and being changed one proceeds by degrees always keeping in view, "better too little than too much." Too little does not hurt the patient, but excess does.

If all danger of a relapse is over then sufficient nourishment may be given to the patient to maintain his convalescence.

Often violent hunger sets in which is, however, a false one, in such a case one must not yield to the patient nor feed him according to his appetite.

It is of frequent occurrence that attendants yield and so cause the patient to relapse into his old illness.

Foresight is the mother of Wisdom.

Day by day the quantity of nourishment given may be increased and by degrees pass into a more solid form.

The convalescent's diet must be the stepping-stone to the diet of the healthy, and in proportion as the digestive organs grow stronger and less sensitive the quantity may also increase.

In inflammation of the intestines and stomach this however is very slowly accomplished; therefore the food must be of a fluid and pappy character until one feels sure that all danger of a relapse is over.

The same caution is needful in dysentery, typhus, peritonitis and enteritis.

The chief nourishment in convalescence should consist of milk, milk soup, mashed potatoes, very tenderly stewed veal, biscuit, poultry, rice milk, strengthening broth and ripe fruit.

As beverages, a little wine and beer may be given to stimulate the digestion.

I have in short paragraphs treated of the food to be given in various diseases. True I have not specified a bill of fare for each individual disease, but I have selected the most important and expatiated on approximate ones.

I do not consider it needful to enlarge on all diseases since in many, which are not feverish, a dietary table is not at all needful.

I have touched on all the febrile diseases and given indications how to act when they occur.

A practical sick attendant or nurse will know how to help under all circumstances and the Doctor decides the diet.

Beyond this it is not difficult if one uses common sense to find out what is suitable for the patient.

In conclusion I would commend to every man's heart that it is possible to prevent some diseases and entirely avert others, if care is taken in days of heatlh to lead a regular life and avoid everything which seems prejudicial to health.

Be moderate in all be what it may, and thou wilt keep thyself in good health.

Cooking Recipes for strengthening the Stomach and for the use of Fever Patients and Convalescents.

Milk-porridge or soup.

Take a pint of milk and put it on to boil with a little salt, keeping back a small portion to mix a quarter of an ounce of flour into a thick soft paste. When the milk boils pour this paste into it stirring it all the time; at the end of a few seconds serve up the porridge. (It is made much in the same way as we in England make arrowroot).

Sweetbread-soup.

Fry a teaspoonful of flour in a quarter of an ounce of fresh butter till it is a pale yellow; then add to it some finely cut onion and parsley and an ounce of skinned calve's sweetbread; pour meat stock over it and cook for about an hour. You may add the yolk of an egg with advantage.

Cream-soup.

Cut up three quarters of an ounce of roll very fine and drop it into boiling gravy and season the whole with parsley, onion, salt. pepper and nutmeg. When the soup has boiled for about half an hour, pour it through a sieve and serve with half a pint of cream beaten up with the yolk of an egg.

Rice-soup.

Scald three quarters of an ounce of rice with boiling water and then boil it with a small piece of butter ($^1/_6$ of an oz.) till it is thoroughly cooked. The yolk of an egg may be added.

Groat or grit soup may be prepared in like manner.

Toasted-bread soup.

Grate half an ounce of toasted bread into gravy, boil it and serve it .up with the yolk of an egg.

One may also serve it with the addition of one ounce of meat which must have been boiled in it for half an hour.

Oatmeal-porridge or soup.

Stir carefully a tablespoonful of oatmeal into some gravy already seasoned and cook it with a little carrot and onion and so send it to table ; a yolk of egg stirred in it makes the soup more tasty.

You can use water in which meat has been cooked salting it slightly.

In the same way you can make soups with vegetables using them instead of oatmeal, or you may prepare
it with lentil-meal, bean-meal or barley-meal.

French-soup.

Dissolve a quarter of an ounce of butter in a frying
pan and simmer in it a little onion and a teaspoonful of
flour till the whole is of a bright yellow. Pour this into
stock already prepared and add very finely cut soup
vegetables such as green peas, potatoes, carrots, etc.
Cook it on a moderately good fire being careful that the
flour does not become lumpy and then rub the whole
through a sieve and add the yolk of an egg.

In the same way also barley-soup and green-corn
soup may be prepared.

Potato-soup.

Boil a quarter of a pound of fresh peeled potatoes
with a little salt, mash them and add a small quantity
of flour and butter and a little milk and if it is not
sufficiently liquid a little meat stock, boil the soup up
just once and rub it through a sieve.

"Luft"-soup.

Stir a dessertspoonful of flour in water until it is
perfectly smooth, then add an egg and during steady
stirring pour in boiling meat stock. Then place it on
the fire and boil up once again before serving.

Bouillon with Additions.

Stir two spoonfuls of flour, two eggs, some milk, salt and nutmeg into a smooth paste and pour it through a coarse colander into the boiling bouillon. It is ready to serve when it boils up well.

Bouillon with Rice.

Place the rice on the fire in cold water; when it boils pour it into a sieve, rinse it well in cold water, put it on again with some bouillon and let it cook till soft, yet so that the rice remains whole.

Water-cress soup.

Having some butter or other fat in the pan, throw in some flour and fry it a regular light brown, this, when thrown into the bouillon or meat stock will be greatly thinned and at the same time bound well together. Simmer for an hour and a half on a slow fire and then take the scum off and drain the soup.

Four or five handfuls of watercresses are now picked over, washed and cooked for a few minutes in salt water. After this drain them, dry them, pass them through a fine sieve, mix them with three or four yolks of eggs and a piece of butter and add to the soup.

Serve it with pieces of roll fried brown in butter and cut into dice.

Chestnut-soup.

Prepare a thin soup with meat stock and white thickening and let it simmer on a slow fire. Meanwhile, remove the outer hard shell of a pound of chestnuts, scald them in hot water and take the inner skin off. After this put the chestnuts into a stewpan with some meat stock and a piece of butter and let them stew gently. When sufficiently cooked pass them through a sieve and mix the paste or purée with the soup from which the fat has been removed: add a small glass of Madeira and pour the whole into the tureen when the soup has been properly seasoned.

Chervil-soup.

Melt a piece of butter, thicken it with one or two spoonsful of flour, add as much bouillon as is necessary for the soup and let it boil up on a slow fire.

In the meantime chop some handsful of fresh chervil very fine, fry it in some butter and pour on to it the well skimmed soup through a fine sieve. Now beat up four or five yolks of eggs with half a pint of cream, clear and strain it and add to the soup with a piece of butter. Serve it with dice-shaped pieces of roll fried brown in butter.

Lentil-soup.

Having well rinsed a pint of lentils, cook them in salt water with onions and vegetables very slowly until they are soft: pour the whole through a sieve into a basin and when the lentils have been drained pass them through the sieve.

Thin the purée or paste with the lentil water and add as much meat stock as is required. After this has been done let the soup boil up once more, carefully skim it, put a piece of good butter into it and serve with small pieces of roll cut into dice and fried brown in butter.

Nudel- or Paste-soup.

Make a firm paste with two whole eggs, two spoons-ful of flour and some salt; roll it out as thin as possible, divide it into strips of an inch wide and then cut them into fine threads, spread them out separately on a table to dry and being dry throw them into boiling bouillon to boil for five minutes.

Ox-tail soup.

Cut into pieces two medium sized oxtails and soak them for two hours: after this boil them up once in salt water and let them cool.

Place them in a stewpan into which you have al-ready put cut onions and other soup vegetables, some slices of raw ham, some pepper corns, a bay leaf and a little thyme.

Let the whole boil slowly in three quarts of bouillon or good stock till the oxtails are soft.

Add a glass of Madeira to the whole.

After this pour the soup through a sieve take off the scum, add a few spoonsful of pearl barley cooked tender, and vegetables cooked in salt water and cut into little dice, season and serve the meat separately.

Sorrel-soup.

Take a pound of Sorrel, pick the leaves from their stalks and chop them fine, together with a little celery and salad and a handful of Chervil.

This being done put them into a stewpan with two ounces of butter and fry for ten minutes; then add as much bouillon or stock as is required and boil slowly.

Beat up four or five yolks of eggs in a basin and whip them well up with half a pint of cream and strain through a sieve into another basin; add two ounces of butter in small pieces and shortly before serving pour the soup over it: it must not be put on the fire again, otherwise the eggs will clot.

Serve with pieces of roll cut into dice and fried brown in butter.

Windsor-soup.

Make a broth of half a chicken and a pound and a half of veal.

As soon as the chicken is tender, take it out and cut the breast into strips an inch and a half long: pound the remaining meat quite fine with some spoonsful of broth, add three or four yolks of eggs and half a pint of cream and rub the purée or paste thus formed through a fine sieve.

In the meantime having prepared some well soaked pearl barley add it to the broth with four ounces of water and some butter and mix with the chicken purée.

Add a glass of Madeira before dishing up and serve with the cut up chicken breast in the soup tureen.

Whole-Meal soup.

The Whole Meal is cooked for a quarter of an hour in boiling meat stock and then served.

Brain-soup.

Having soaked Calves' brains in water, mix them with a finely chopped onion and fry them in butter, dust them with a little flour and fry again; stir them into some meat stock and boil for a few minutes.

Serve the soup in a tureen into which toasted slices of bread and yolks of egg have been placed.

Liver-soup.

This is made in a similar manner to brain soup with the exception that it is strained through a sieve before serving.

Sweetbread-soup.

After cleaning and skinning the sweetbread pound it in a mortar and prepare it like brain soup.

Brenn-soup

or what we call Poor Man's broth.

Take a little flour and a finely cut up onion and fry them in butter or dripping; then add water and season the whole with salt and nutmeg and serve with toasted slices of bread and a beaten up egg.

Black Bread soup with Egg.

Take some thin slices of rye bread, steam them till soft in hot butter, pour salted water slowly over them and having beaten them up cook in water to a thin paste.

Have ready three yolks of egg beaten up with cream and a little seasoning and mix all together just before serving.

Cheese-soup.

Grate ten ounces of Swiss cheese and place it in layers in the soup tureen with alternate layers of slices of bread.

Add to a little brown thickening some finely choppep onions, seasoning, and a quart or three pints of water; boil all up together and pour it on the cheese and bread in the soup tureen.

This may stand for some little time on the stove to give opportunity for soaking before serving.

French Beans or Scarlet Runners.

The beans are strung or better still peeled on both sides: next cut them in thin strips and cook them till very tender in salt water so that they may remain nice and green.

It is not at all necessary to cook them in a copper pan, any other serves the purpose quite as well.

The principal things are plenty of water and rapid cooking.

When the beans are soft throw them on to a colander and cool them directly with cold water until they are completely cold.

As soon as the beans have been well drained put them into a shallow stewpan, well greased with butter, dust them with pepper and stir them continually until they are quite hot.

Add some chopped parsley and serve.

Duchess Potatoes.

Take about two pounds of floury potatoes, peel, wash and cook them till soft in salt water.

Then drain them quite dry and let them remain open before the fire for a few minutes to get rid of the steam and rub them through a sieve.

Place this purée or paste into a pan, mix it with three or four yolks of eggs, two ounces of butter, peppei and nutmeg and being well mixed let it cool off.

The whole is placed on a table strewn with flour and rolled out, then cut into pieces the length of a finger: press these flat with the blade of the knife and you have elongated squares of about half an inch thick.

Heat some clarified butter in an omelet pan, lay the potatoes in it and fry on both sides to a nice golden brown colour.

Potato purée.

Choose very floury potatoes because the watery kind become tough in the process.

The potatoes are peeled, washed and cooked tender in salt water. Drain them dry and let the steam pass off a minute, put in a piece of butter and let it soak through the potatoes; then rub them as fast as possible through a hair sieve and put them into a stewpan and stir them until they are foamy in hot milk or cream.

Finally you put yet another piece of butter in and serve after well stirring it.

Sorrel.

Take a few pounds of young sorrel, strip the leaves from their stems, wash them thoroughly and cook them in salt water till they are tender: then drain them and pass them through a sieve and boil again quickly with some butter until the whole has become a thick paste or purée: add salt, pepper and a little sugar and just before serving a piece of fresh butter.

Serve it in a dish and lay round it poached eggs.

Fried Potatoes.

Boil the potatoes, peel them, cut them small and fry them in butter with onion, salt and pepper.

Mashed or grated Potatoes, fried.

Cold boiled potatoes are grated or mashed, fried in
a pan with fat and onion until they are gold colour:
serve loosely on a dish.

Tossed Potatoes.

Peel raw potatoes and cut them in slices: put them
into a buttered mould with salt and a piece of butter
and bake for an hour in the oven: turn them out and
serve.

Turnips with Potatoes.

Peel some turnips, cut them in long strips, place
them in boiling water with salt and cook them till tender.

Then raw potatoes are peeled, cut in slices and added
to them.

Some flour fried yellow in butter is then mixed with
the potatoes and turnips and all cooked together till soft.

Carrots.

These are scraped, washed and cut; then steamed
in fat and dusted with flour; add meat stock, a little
sugar, and salt and steam until completely soft.

Dandelion vegetable.

The hard points being removed the small leaves are
washed, cooked for an hour in boiling water, then strained
and chopped very fine.

Put these prepared leaves into a stewpan with a
piece of butter and some salt and steam them for a

while; then lightly dusting with flour, add water and cook altogether for a quarter of an hour.

Carrots with Potatoes.

Put the carrots into slightly salted boiling water; after they have cooked some time, slices of raw potatoes cut lengthwise are added to them. Cook the whole extremely tender, stir into the mixture a spoonful of flour with water or meat stock, and serve it with onions fried golden brown.

Bavarian Turnips.

These are peeled and the larger ones cut through once, the smaller ones left whole: wash them and put them into a stewpan with sugar and butter or dripping; moisten them often with meat stock. Steam them till they are soft, dust them with flour, season and shake them well together and serve.

Black Hellebores.

After these have been carefully cleansed they are cooked in salt water with milk and lemon juice; when the water is poured off strew with grated bread and cover with melted butter.

Savoy Cabbage.

The cabbage being well cleansed it is cooked till tender in salt water, shaken in a sieve and well drained: it is then cooled in cold water, put into a pale yellow glaze, steamed and served with meat stock and salted.

Baked and Stuffed Savoy Cabbage.

Cook the well cleansed heads tender in salt water and carefully take off the leaves: chop half a pound of raw veal and pork very fine, stir some milk into it and add two rolls well squeezed from the milk in which they have been soaked, together with four eggs, salt and pepper: mix all well together.

Next a mould is buttered; the savoy leaves are laid double on the bottom, then a layer of forcemeat, then leaves again, then forcemeat till finally some leaves are laid on the top with pieces of butter.

Bake in the oven for an hour.

Sauerkraut (sour-crout).

Boil this in water till tender, make a flavouring of burnt onions, stir it into the crout and, when it has boiled, serve.

Ribwort Plantain & Stinging Nettle as vegetables.

These being well washed are cooked tender in water and chopped fine with onions. The mixture is then steamed in clarified butter mixed, with meat stock, salted and seasoned and then for a time allowed to boil.

Spinach in Omelettes.

Make an omelette: chop some spinach fine and mix with it some few spoonfuls of melted butter, two eggs, a little salt and bread crumbs; brush the omelette with it and put it into a buttered stewpan, pour meat stock over it and cook it in the oven for half an hour.

Asparagus.

Scrape the asparagus and so remove the stringy part. Cook in salt water and serve with a butter sauce according to taste.

Green Peas with Carrots.

These are steamed gently in butter till soft, dusted with flour, moistened with meat stock, mixed well together, salted and sent to table.

Green-Peas.

Having removed the shells steam them in butter till tender, shake a little flour over them and moisten with meat stock.

Vegetable Ragout.

Haricot beans, green peas, square cut potatoes, soup vegetables and radishes are each separately well cooked: then with the exception of the radishes all are drained. the next process is to mix some flour with sour cream and the radish water and boil it. Shake into this mixture all the vegetables which have been prepared, mix well together and then let it get cold.

Lemon juice may be added.

Oatmeal-Porridge.

Pour three ounces of grits into a quart of boiling milk with some sugar and boil for at least an hour stirring it constantly.

Flour-Milk-Porridge.

Three ounces of flour are stirred with water into a thick pap or pulp; this is poured into a pint of boiling milk sweetened with sugar and cooked. Rice-flour-Porridge is made in the same way.

Rice-Milk-Porridge.

Scald three ounces of rice two or three times by pouring boiling water over it: then place the rice with some sugar into a quart of milk; cover it up and let it cook slowly until it is swollen and thick.

Sugar and cinnamon may be eaten with it.

Vegetable-Porridge
(Bean flour, Lentil Meal and Pea flour).

A pint of meat stock is brought to the boil: into this you throw the meal you intend to use and when it has become pulp or pap, add some butter and cook for a quarter of an hour.

Mashed Potatoes.

Peel a pound of raw potatoes and cook them in salt water till tender, drain them through a colander and mash with half an ounce of butter and boil for a few minutes in a quarter of a pint of milk. The pulp may, if desired, be passed through a sieve before serving.

Soft-Boiled Eggs.

Place the eggs in boiling water and cook them from two to four minutes.

Pancakes.

Into ten ounces of flour stir gradually five eggs and a pint of milk until it presents a smooth paste. Season with a little salt. Dip some of this out into a pan with hot butter and, while frying, cut the mass into small pieces.

Omelette.

Mix three eggs with three quarters of an ounce of flour, some milk and a little salt. Fry the omelette on both sides and send it covered to table.

Boiled-Beef.

If you want to obtain a good meat stock, put the meat on in cold water with vegetables and a little salt and cook till tender.

If, however, you wish to serve the meat as boiled beef, it is put on in boiling water which makes it more tender. In the process of cooking take the saucepan off the fire for a time and then while it is still warm, set it on again and cook it until completely tender.

Beef with Sour Cream.

Take a piece of rump of beef, wash it, beat it, season it and let it stand: set it on the fire with a pint of water and a pint of wine with onions, soup vegetables, cloves, lemon peel and a bay leaf. Cover it well up and steam it for two or three hours. In the meantime prepare a thick brown flavouring and mix it with a little cold water and some of the liquor in which the meat has been steamed: let this sauce cook till it is thick, then add half a pint of sour cream and cook it for another ten minutes. Pour it over the meat and serve.

6*

Sirloin or Fillet.

The joint is washed, salted and seasoned and placed in a pan with onions, carrots, turnips and bacon together with some water in the oven. The joint must be constantly basted and cooked until it is of a nice brown colour and tender.

The sauce is served separately.

„Gulasch".

Cut part of a breast or neck of veal into joints, salt and pepper them. Make some butter or fat hot in a frying pan and lay the pieces of meat in it with a finely cut onion and fry quickly in order that they do not get hard.

"Gulasch" can be made of beef as well as of veal

Roast Pigeons.

These are prepared in the same way as chicken; they are then stuffed with seasoning and roasted a nice golden brown.

Roast Hare.

After the hare has been cleaned, washed and larded with bacon it is seasoned with salt, pepper, cloves, cinnamon, onions, carrots and bay leaves: then vinegar and wine are poured over it and it is left all night; then it is covered up in a stewpan and steamed, after which it is roasted.

Should the hare be an old one it must be left longer in the pickle.

English Mince or Rissoles.

Half a pound of beef or veal and a pound and a half of pork are minced fine with onion and lemon peel, salted and seasoned: then add four eggs and a few spoonsful of bread crumbs and make the mixture into little cakes or rolls: cook them in a stewpan with a piece of butter and put them into the oven to brown, basting them from time to time with meat stock or gravy.

Stewed Veal.

A piece of veal is washed, rubbed with salt and pepper, put in a pan with butter or fat and allowed gently to stew with the addition of onions and carrots. A little sweet milk poured in until the meat is soft is an improvement and makes a beautiful sauce for the meat.

Stuffed Breast of Veal.

A boned breast is salted, seasoned and stuffed with the following stuffing: Four rolls are soaked in water, well squeezed out and steamed in butter with onion and parsley: add five eggs, mix it all together and stuff the breast. Stew the latter in butter.

Roast Lamb or Kid.

The meat is cut in quarters rubbed with salt and pepper and roasted in butter in the oven: during this process pour some water over it and baste it with the gravy or sauce. It will be ready in a quarter of an hour.

Third Part.

Gymnastic Exercises.

(Home Calisthenics.)

General.

It is a well known proverb that "in a healthy body there lives a healthy mind".

If this be true, and I do not in the least doubt it, then it is doubly incumbent upon us to keep our body strong and healthy. It is quite true that if the body be ill and sickly the mind suffers with it by the pressure exercised by sickness on our nature.

The illness need not be particularly severe in order to produce mental depression.

It is said one is simply not in good spirits, yet the reason for this lies in the bodily illness of the man.

We know by this time the varied curative methods and their application and I believe that the advice I have given ought to suffice.

Yet I would still gladly speak upon a point about which there is a good deal of talk, that is Gymnastics.

Our forefathers practised Gymnastics in very early times and we read that the ancient Teutons and Romans employed this kind of exercise for the purpose of strengthening their muscles.

Even the children were accustomed to these exercises and attained a certain facility and agility in them. In fact work strengthens the muscles and it is a matter of daily observation that men who have never done hard work have rarely any muscular power. For officials, handworkers, clerks, professional men and the like who never get the opportunity of strengthening their muscles, I think Home Gymnastic Exercises very good; at the same time I do not approve of difficult exercises and for anyone suffering from heart disease in any form I should strictly forbid them.

Everything has its dark and bright side and Gymnastics form no exception.

Many men overdo everything they try; of this I have often convinced myself: they believe that the more they do the better it is: this however is wrong. Everything must be done in moderation and with method if it is to act beneficially on our system.

Do not be in a hurry for results, everything worth having must be steadily worked for with patience and perseverance.

Gymnastics are not only of service to healthy people, but to invalids also if used in the form of curatives.

The skeleton is the foundation of the human body, it supports the soft parts and protects its various organs, as for example the chest and skull.

The muscles are attached to the bones and exercise upon them an influence by which movement is produced: the stronger the muscles the more supple the joints and the easier the movement.

It frequently happens that for lack of exercise certain joints become stiff and lose their suppleness. In such cases Gymnastics are of service unless of course the joint has become useless and deformed.

The muscles require nourishment and this is supplied by the blood whose duty it is to nourish all parts of the human body.

The nutritive ingredients of the blood consist of albumen, fibrine, salts, fat and water which are the actual building materials out of which the tissue of the body is formed.

As it is necessary that the building materials should not fall short, they must be added to in the form of nutritive matter which we imbibe. These are equivalent to the raw material from which the building stones or whatever else they are called, are formed: and here again I emphasize the fact, that the more strengthening our nourishment the better is the material for building up our system.

All living things are subject to a continual interchange of "stuff" or material; that which is unusable is rejected while that which is good is employed by the system : this happens every day.

Every day therefore we must eat and drink in order to supply the body with new material.

Is it possible for a man to strengthen his muscles without movement ?

Certainly not.

Everyone knows by experience that after energetic movement the heart beats stronger and quicker, the pulse is more animated, the internal warmth increases and as a consequence of this the appetite is healthier.

The reason of all this is that the movement caused the interchange of matter to be more marked and the existing material to be used up more quickly. Therefore people who work hard have as a rule a much better appetite than those who live a sedentary life. The muscles being active cause a more active interchange of material, and bring about a desire for food.

In the same way a suitable nourishment of the entire organism acts beneficially on the nerve system which forms part of the whole.

If the nerve system acts in a normal way and can supply itself with proper nourishment, the gratifying result will be that after the daily exertions of the body there will come a quiet health-giving sleep, a sign that the whole organism is in harmony.

It is with men as in clock work; the wheels are all related to and depend on each other and if the smallest fails in its duty the whole machinery is brought to a stand still.

Nothing more perfect or more beautifully harmonious ever came from the hand of the Creator than man. Therefore Man thou art bound to offer praise to thy Creator.

Everything about man has its signification and its own appointed function.

If therefore but a single part of the body is hindered in its full activity, the whole man not unfrequently suffers.

Unfortunately the individual himself is often to blame for the disturbance in his organism and must bear the consequences which he has brought on himself.

One should not wait for illness, but look after the body in health taking regular diet and physical exercise. Even an infant requires exercise, it is, as it were, born to it: it makes movements, kicks and crawls and in this way creates exercise tor itself.

One should employ the utmost care in dealing with a child, first that it is scrupulously clean, next, that in lying down it can freely move its body, which on no account must be cramped.

The child's clothing must not be at all tight, it has an impulse to throw off every thing that is irksome and if it cannot free itself it makes known its discontent by loud crying.

It is a mistake to force a child to sit, stand or walk before it has the strength to do it. One must wait until it feels its way to do these things. It will as soon as it is able try to sit up and involuntarily make an effort to stand: then be ready to give it support so as to make the experiments easier.

Premature sitting up often results in spinal curvature, even the exclusive holding up of the child on one and the same arm is harmful.

As soon as the child has learnt to step, its impulse is to play and tumble about, it seizes whatever it can lay its hands on to play with : it wants to work without knowing why: it is simply impelled of itself to strengthen its body, to harden and exercise it.

The Proverb runs : "Let children sow their wild oats".

I understand this in a reasonable sense viz., that one should give them opportunity by exercise and suitable play and gymnastics to drill themselves into a healthy condition, not however so far that they degenerate into wildness and arouse the passions in them, for these must be crushed in childhood and never allowed to break out into flame.

What I especially wish to emphasize is that children should never wear dress that cramps them whether the article be shoes, trousers or dresses.

The perspiration of the body should never be hindered and its development can only be promoted by wearing loose articles of dress.

Avoid rigidly all lacing in whether by garters or by waistbands.

The shoes should be broad and not follow too exactly the shape of the foot which should have room to develop. The heels should never be high but on the contrary broad and low and the toes should not be cramped.

Such measures will sensibly promote the development and remove obstacles to healthy movement.

In the country one generally lets children run very freely about so that the advice to dress them in loose garments is not so necessary.

Children in towns have to strut along prettily and the mothers are careful that the clothing fits their bodies to a hair's breadth so that they may appear dainty and finical.

It is the same as regards bodily exercise: children in the country are up betimes, almost too early indeed, and go to work which they are bound to learn and thus their bodies are strengthened and their limbs made supple.

This is certainly the best form of gymnastics and with the addition of sufficient strengthening food the entire system is developed in the best possible way.

With town folks this is different: they have as a rule far too little free exercise and too little bodily labour, add to this in many instances the want of fresh air and the lack of proper nourishment.

In large cities the walk to school and back is often the only exercise the children get, that is if their schools give no instructions in Calisthenics.

The air in large cities is anything but inviting and the smoke from factory chimneys is a large contributor to lung disease.

In these cases Gymnastics are very good and home Calisthenics are to be recommended.

Girls should strengthen themselves by appropriate exercises, all the more because their work in the future is likely to be sedentary, which, if not corrected, produces an ill developed body and an insufficiency of blood.

Very much weight must be laid upon strengthening the chest; the lungs must be able to expand and air to penetrate the most minute air cells. **Therefore no corsets!**

As the child develops into youth I consider that individual exercises are in place.

Such people as are inclined to obesity or possess elastic muscles may undertake strong exercises. Older people may acquire strength by regular movements and exercises suitable for their age.

People accustomed to Gymnastics in their youth will accomplish the exercises more easily when they are old than those who were formerly stinted and starved by physical inactivity.

Now, what exercises can one really carry out?

Every period of life has its own, the child, the youth, the maiden, the man, the woman, the greybeard and the old lady.

All do not demand the same strength, and we have to reckon with weak and strong, healthy and sick people.

I will now give a few indications as to which appear to me the most useful and the easiest to carry out.

Representation of Individual Exercises.

Movements of the Head.

The movements of the neck including those of the head are to be carried out quietly.

Turning the Head.

Standing perfectly upright and without moving the shoulders you turn the head as far as possible first to the right and then to the left until you look back over the shoulder.

This exercise is good for stiff necks which it gradually renders suppler.

Bending the Head Sideways.

This bending includes backward and forward and sideways right and left.

Turning the head.

Bending the head.

In the forward movement the head should be bent so low that the chin easily touches the breast and the backward movement should be carried out as far as possible.

It is not necessary to exercise great force, but the shoulders in the meantime must be kept quite quiet.

For the sideway movements the position is upright: one bends the head left and right without turning it or raising the shoulders.

Exercises for the Body.

In Paralysis or lameness which precludes activity of the muscles gymnastics are useless; in partial paralysis the exercises may be tried with some success.

I recommend invalids not to use Gymnastics except under the supervision of a doctor or an expert.

Turning the Body.

With the feet standing close together flat on the ground and the figure perfectly upright you cause the upper part of the body to turn to the right and to the left with the hands held out straight in front and without moving the legs at all.

Turning the body.   Bending the body backward and forward.

Bending the Body backward and forward and to left and right.

Just as the head can be turned in various directions so also can the body.

The hands are supported on the hips, the legs remain quiet, the heels are placed close together while the toes including the feet take an outward position: the shoulders must be kept quite still during the movements backward, forward and sideways.

The exercises must be performed in moderate time and all haste avoided.

In the backward movement draw in a long breath and in stretching oneself breathe it out again. In the forward movement one breathes out and in straightening oneself breathes in.

Bending the body sideways.   Bending the body while turning it.

These exercises are of great benefit to people suffering from piles, constipation, weakness of the abdomen and weakness of the muscles of the spine.

They have also been of great service in temporary curvature of the spine, yet the bending must only be made on one side according to the direction of the curvature.

Exercises for the Arm and Hand.

If one has a stiff wrist which is not an old standing affair, use the strong hand for the purpose of bending the stiff wrist backward and forward. As a support put the hand and wrist into a warm hand bath and subsequently into a cold douche, and after drying the hand, start the exercise which must be gradual, that is a little more bending every day, for such stiffness is not removed without difficulty and perseverance. A rotatory movement may form one of the exercises.

After taking off a plaster of Paris or other stiff bandage by which the joints are forced to keep in one and the same position for months, a stiffening of the joint very frequently occurs whether it be in the arm or leg. By these exercises then these disagreeable circumstances may be removed.

In weakness of the arm or hand these exercises are of the greatest benefit.

Raising the Shoulders.

In weakness of the muscles of the shoulders the so called shrugging of the shoulders is good. This exercise is not difficult and offers many advantages for activity of respiration of the lungs.

I recommend specially to persons with diseased lungs the practice of drawing deep breaths by means of which air is pumped into the innermost parts of the lungs and necessitates expansion of the ribs.

The raising of the shoulders must be done with the arms hanging down: the former should be raised slowly and with as much power and as high as possible, and again lowered slowly; this exercise may be carried out with one or both shoulders at the same time, or first with one and then the other.

Draw in a breath while raising and give it out while lowering the shoulders. If one is high and on one side

7*

raise only the lower one. In moving the shoulders backward and forward support your hands on your hips and move the elbows backward and forward, the upper part of the body remaining upright.

Moving the Arms forward and sideways.

I beg you to observe that in raising the arms you take a breath in and breathe it out when lowering them. Raising the arms sideways is done in this manner: the fully outstretched arms are slowly moved to the side till they are perfectly horizontal, then they are raised into a perpendicular position, whereby both arms remain apart from each other the width of the body and the palms of the hands turned towards each other.

These exercises expand the chest and improve the breathing powers.

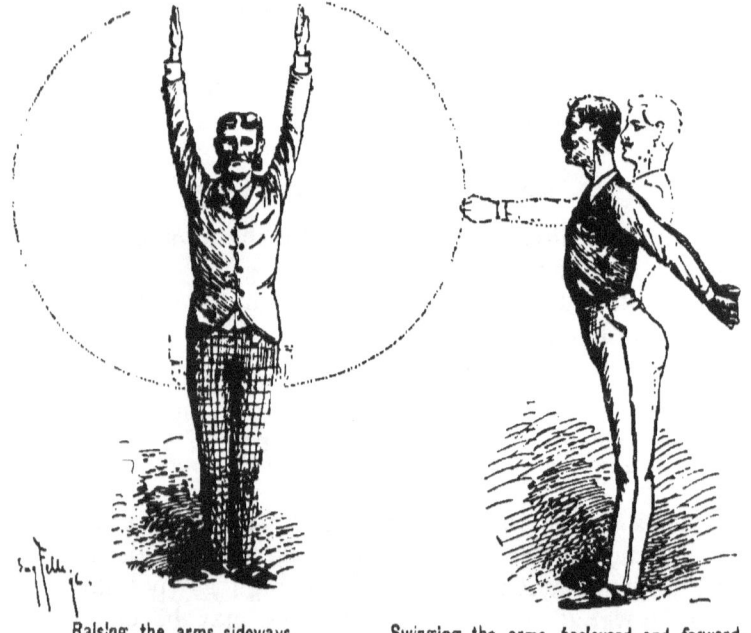

Raising the arms sideways.   Swinging the arms backward and forward.

Swinging the Arms.

A further kind of Arm Exercise is the swinging of them sideways and backward and forward. The picture above illustrates the way to do this. It is not a difficult exercise. The hands are clenched and raised into a horizontal position. In swinging them sideways swing them outwards into a perpendicular position and let them fall rapidly on to the legs again. Take in breath while raising them and breathe out while lowering them.

In swinging the arms backwards and forwards get them as far as possible behind. (See illustration.)

This exercise may also be done with dumb bells provided they are not too heavy.

The first picture below shows us the Arm Circle in its conic form, also that the arms being raised horizontally and stretched out a circular movement is made without the arms being at all contracted.

Conic circle.   Large circle.

A somewhat similar exercise gives another species of Circle viz., the large arm circle called Mill Circle in which the arms describe big arches beginning either forward or backward.

Same with dumb-bells.

The illustration shows the commencement and the finish.

The inspiration and expiration of the breath is carried out as in the former exercise.

In cases of malformation of the chest, or in holding oneself badly or in curvature of the spine these exercises will do good service.

Bending and Stretching of the Hand and Hand Circle.

These exercises as well as bending and stretching the fingers are beneficial in Writer's cramp, in convulsions like St. Vitus' dance and after fatigueing activity of the hands or stiffness of the joints.

Bending and stretching the hand with or without dumb-bells.

Bending and stretching the arms.

The exercise is very easy: the bending and stretching of the hands may also be made with dumb-bells either upwards, downwards or sideways.

The movement upward is made with outstretched arms and clenched fists and the fist brought back to its first position. The sideways bending is done in this way: the outstretched hand being bent towards the thumb or the little finger. Finger bending is neither more nor less than clenching your hand and opening it again.

As you bend and stretch the hand, so may you exercise the arms. As in the exercises previously shown the body is maintained in an upright position. The arms are so bent that the tips of the fingers touch the shoulders in front.

Starting with them in a bent position you stretch your arms out in front, behind and sideways, up and down, just as is shown in the second illustration above.

The stretching may also be carried out on one side only that is to say merely with one arm, the right or the left.

The arms may also be stretched down backwards, and in fact in the following way, but I again caution you that expiration must correspond with the stretching, and inspiration with the bending.

(a)　　　Stretching the arms down behind.　　　(b)

The hands are folded at the back (a) then the arms are slowly and forcibly stretched down and the shoulders drawn back and down (b).

These exercises are of great benefit in bad deportment, crooked backs, asthma and difficulty of breathing.

Thrusting with the Arm.

This thrusting with the arm and stroke with the fore-arm are somewhat similar in character to those above.

Thrusting the arms
downwards.

In thrusting the arm the hands are clenched and by a forcible push stretched forward, sideways, upwards and downwards.

Just as in stretching the arms, so in thrusting them, the exercise may be performed with the right or left arm.

Thrusting alternate arms.

The fore-arm stroke can be done in two ways, with wrist position and raised or crest position, the former when the back of the hand is turned up, the latter when it is turned down.

Fore-arm exercise with wrist position.   Fore-arm exercise with crest position.

During the sideway movement of the arms a deep breath is drawn in, and during the forward movements it is given out again.

During the practice of the fore-arm exercise in the wrist position, the arms are stretched out horizontally, the arms are bent inwards so that the finger tips touch each other in front of the chest; afterwards the arms are swung or outstretched slowly, but always in ho-

Thrusts with alternate fore-arm,

rizontal position. In the second position the arms are stretched out as in the illustration, the fore-arm being bent so that the finger tips touch the shoulders.

The upper part of the arm remains in a horizontal position while the movement of the fore-arm proceeds.

The exercise may be performed with a swing or slowly. The effect of this exercise is similar to those already given.

Foot and Leg Exercises.

Just as we can raise the arm so can we raise the leg, the exercise being similar.

Raising the Leg.

This can be done sideways, forwards, backwards, as well as obliquely backward and forward.

With the hands supported on the hips the right or left leg is raised slowly sideways, forwards or backwards, as well as according to the direction required.

Raising the leg sideways.   Swinging the leg backwards and orwards.

The upper part of the body must be kept perfectly quiet; therefore you can only raise the leg so high as will permit of the body maintaining a perpendicular position.

The leg on which you stand must not be bent, but the point of the foot should incline outwards so as to make it easier to stand.

Keep the elevated leg in that position for a short time and then draw it slowly back.

Swinging the Leg.

These exercises include swinging the leg forwards, backwards, inwards and outwards.

The foot is pointed outwards; you raise your body somewhat, so that in swinging the outstretched leg backwards or inwards or outwards you do not touch the ground with it.

If you cannot accomplish the exercise unaided use a chair as a support. The elevation of the leg is done slowly, but the swinging it is accomplished with energy and rapidity; in this lies the difference between the exercises.

Circular Motion of the Leg.

By this is understood a movement of the leg in such a manner that either the right or left leg is raised in front, and the foot with an equable movement is moved outwards, behind, inwards, and again passed back in front of the standing leg, so that the movement forms a circle.

In stiffness of the hip joint and for constipation these exercises are greatly to be recommended. They also invigorate and strengthen

Outward and inwards swinging of the leg; circular movement of the leg.

the muscles. It frequently happens that an illness leaves behind stiffness of the joints of the toes, and of the knees; in such cases it would be quite worth while to try the following exercises.

They will prove interesting and easy to practice. For those people who cannot well stand, a chair may be used as a support, otherwise they should be carried out unaided.

Bending of the Knee.

With the hands resting on the hips the knees are quietly and equally bent in the direction of the feet until the thigh and the calf of the leg form a right angle. In this movement it is necessary to raise the heels until one stands on one's toes; the upper part of the body remains in a quiet upright position. This exercise must be practised slowly.

Slight bending of the knee.

The deep bending of the knee is like the former with this difference, that the genuflexion is made as low as possible, so that one almost sits on the ground. As in the previous exer-

Dee bending of the knee.

Alternate bending of the knee.

cise the heels are raised and remain close together. The upper part of the body remains in a perpendicular position and must not be bent forward.

These exercises are to be recommended in cases of partial crippling of the extremities: they may also be tried in stiffness of the knee-joint or ankle.

The position for the alternate bending of the knee is not difficult. One leg is advanced a couple of feet and bent so that one knee is inclined towards the ground while the other is pointed upwards. This is quite an easy movement made almost in the way in which one bows.

Raising the Knee.

Raising the knee is not really difficult, but it necessitates standing on one leg which some people cannot manage without support.

The knee is raised as high as possible and then the leg is stretched out: in this movement the thigh becomes horizontal of its own accord and it must not be lowered further than the stretching requires.

Elevating the knee and stretching the leg forward,   Raising the lower part of the leg.

In similar fashion, the raising of the lower part of the leg is accomplished. The lower part of the leg is bent behind as far as possible so that the heel approaches very near to the body.

This exercise can be done with a swing or slowly, always taking care that the rest of the body remains in a quiet upright position.

Exercises with the Pole or Staff.

One can practice a variety of exercises with a pole; naturally in these the arms are most occupied though now and again the legs take part.

The pole used is generally of wood, except a very strong person wishes it of iron: its length should be a yard and a half.

These exercises are more for the healthy than for invalids except the latter be merely weak from poverty of blood or as a result of former illness.

In beginning the exercises hold the staff horizontally before you with arms stretched downwards: the hands are equally distant from the body and so turned that the back of the hand is towards the front. Starting from this position the staff is raised, lowered, or swayed.

.Holding the staff in front of the body.   Raising the staff above the head.

The staff which as indicated was held horizontally in front of you, is now raised in front until the arms are horizontal and from thence to above the head : After a time lower the staff back to its first position. Take a deep breath while raising it and give it out when lowering.

Lowering the staff behind has the foregoing exercise as a foundation, seeing that the staff when held above the head is lowered behind instead of in front; in fact the arms are simultaneously bent until the staff touches the shoulders; from this position the arms are stretched downwards behind and in the same way back again including the front position.

Lowering the staff behind.

Until one acquires dexterity hold the staff at full length in order to make the backward bending of the arms easier to carry out.

Raising the Staff Behind.

The next illustration shows how to raise the staff high with both hands from behind. The difference consists in holding the staff low down behind the body without any movement of the arms.

With outstretched arms you raise the staff above the head (it may also be swung up) and then lower it behind without bending the arms.

How the sideways swinging of the staff is done the illustration shows clearly.

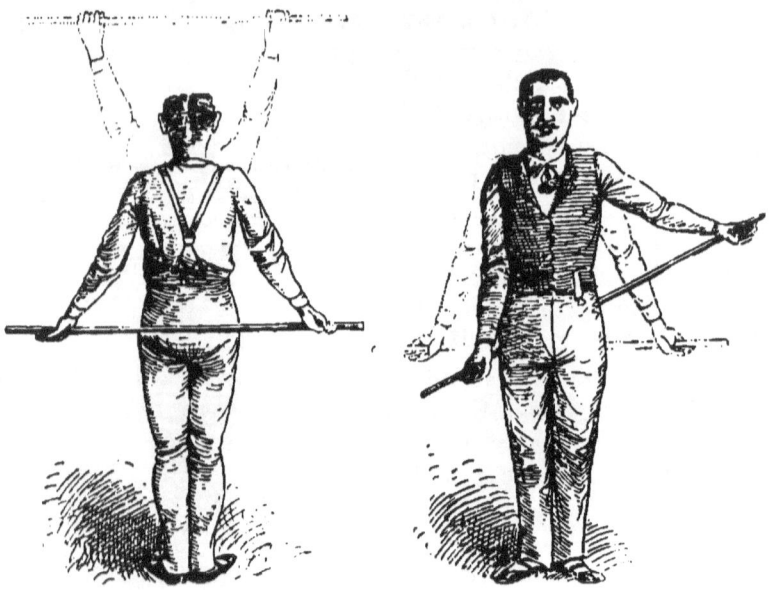

Raising the staff with both hands simultaneously.　　Swinging the staff sideways.

Lowering the staff sideways.

Lowering the Staff Sideways.

From raising the staff above one's head one passes at once to the sideways lowering of the staff by stretching the left or right arm upwards at the side till the fore-arm gets over the head as nearly as possible in a horizontal position while the body remains in a perpendicular position. The exercise may also be done with a swing. Bending the head or the body must be avoided.

Stretching the Arms Sideways with the Staff.

As in the exercises already shown the staff is held in front and the arms are stretched forward. From this position the present exercise starts: the staff is raised as high as the shoulders, bending the arms slightly; then both arms are energetically stretched sideways.

Stretching the arms sideways in front,   Stretching them sideways behind,

At the same time the right arm, which has remained bent, is with the left, raised as high as the shoulder following the movement with it.

This exercise can be practised either in front or behind as shown in the illustration. The staff is raised high above the head and the arms are bent so far that the staff touches the shoulder.

Starting from this position you can exercise the arms alternately.

Rocking the staff.

The staff is held horizontally in front and swung by the right arm into the perpendicular while the left arm bends itself.

Kneipp, Codicil,   8

Rocking the staff.

Starting from this position the rocking of the staff may be done to the left, causing the left arm to raise it to the perpendicular while the right arm is bent.

This proceeding is repeated several times to right and left. The body meanwhile must not be bent nor may the hips and shoulders be pushed out.

Swinging the Staff Sideways with Turning of the Body.

The staff being held horizontally in front is swung with outstretched arms obliquely upwards alternately right and left, the body at the same time turning to the side towards which the staff is swung. The feet in the meantime must remain firm and close together, not turning with the body.

Mounting the Staff.

As seen in the illustration close by, the staff is held horizontally in front of you and the arms are stretched downwards; the backs of the hands being turned to the front. The right or left foot now mounts the staff between the two hands

Mounting the staff.

without bending the body, thereby raising the knee forcibly.

From this exercise you can pass on to that of clearing the staff with the left or right foot without touching it and without having moved the staff from its former position: the knee is raised quickly and the toe of the foot is placed lightly on the ground.

After this follows the return mounting of the left or right foot always taking care to avoid bending the body.

All these exercises have great influence ou our muscles and the various organs of our bodies and they are useful in youth removing many defects in the position of the body and correcting malformation of the chest.

There are still two more exercises in connection with the staff: the first is walking erect with the staff behind. The staff is laid straight across the back and held with both arms bent; the hands held towards the front are clenched, and then you are ready for the exercise. The body assumes a slight bend forwards, the shoulders are drawn back and the chest well thrown out. Then walk forward taking long steps. The muscles of the legs must be forcibly exerted and on putting the foot down, the toes must touch the ground first.

Walking erect with the staff held behind.

There is yet another exercise of like character viz., you place the right or left foot a few paces in front, bending the advanced leg so far as to bring the knee perpendicularly over the point of the foot.

8*

Marching with the staff behind.

The upper part of the body is now inclined forward and comes into the position indicated in the accompanying illustration with the hind leg stretched out.

The feet are placed flat on the ground while the position of the staff is the same as in the preceding exercise. It is practised alternately with the left and right leg.

Swinging Exercises.

Many people especially those occupied in offices try to exercise themselves by taking up wood-sawing and chopping; but then everyone has not the opportunity of doing this and it is to these I recommend the following exercises which seem to me eminently practical.

Both in sawing and chopping wood the arms make a movement while the rest of the body is almost tranquil. The exercises I propose to take the place of these movements I call by their names viz., Ist, **Felling with an axe.**

As the illustration shows, the legs are placed somewhat sideways apart, the arms are raised high perpendicularly and the hands clenched.

Felling with an axe.

Starting from this position the arms are swung briskly down while the body bends simultaneously.

In bending forward and by the impetus of the swing the arms go near the knees and bend themselves a little, while in swinging the arms and body up, the head and the body bend slightly backward.

Again another exercise the movement of which is like that made by a countryman when mowing. The arms are raised in front horizontally and swung sideways. The right arm in swinging remains outstretched and the left bends itself at the elbow-joint in front of the chest and vice versa.

The body inclines somewhat forward by a slight turning of the hip joint and, without the head being moved from its proper position, follows the swing of the arms more or less.

Rotatory swing.

It seems to me that the exercises I have briefly, and I think intelligibly delineated are quite sufficiently useful as I do not care for difficult ones to be practised.

My desire has been in these exercises to give simply some directions to weakly people and those who lack physical exercise as to how they may brace and strengthen their bodies by Gymnastics.

Invalids may practice these exercises as far as they are suitable for them: yet I advise invalids before making use of any of them to take a Doctor's opinion.

As arm and chest expanders I would recommend for Gymnastic exercises the "Patent Largiadère": how these

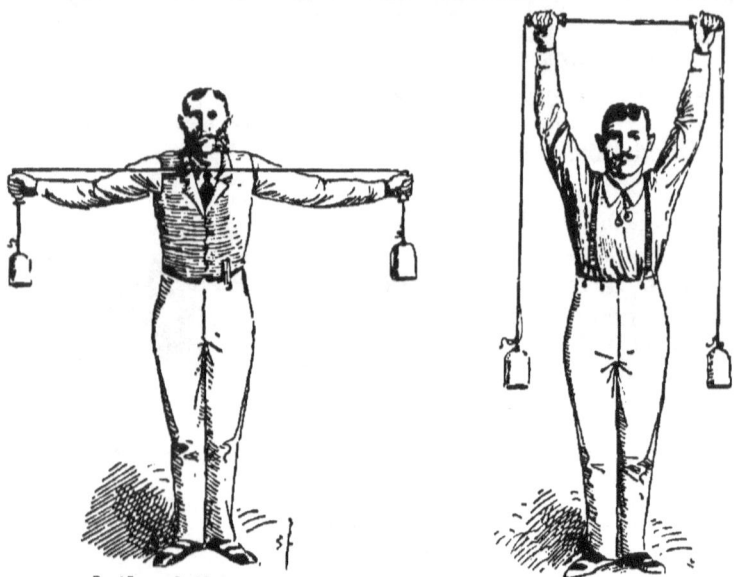

are used the following four illustrations show so clearly that any explanation would be superfluous.

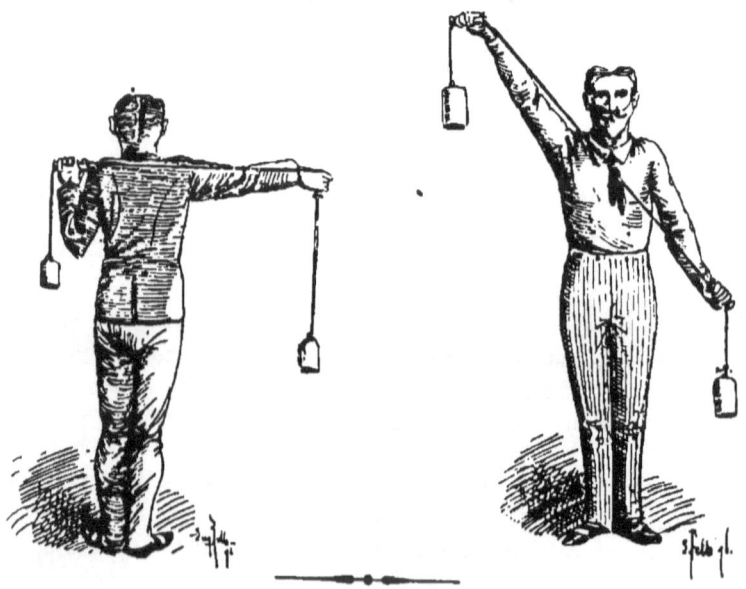

Fourth Part.

Formation and Care of the Human Body.

Formation of the Human Body, the Action and Purpose of its Organs.

General.

The body consists of hard, soft and liquid constituent parts.

To the first belong the bones, cartilage and teeth; these, together with the compact system of ligaments by which they are bound together, form the **skeleton**.

The skin, the fatty tissues, the muscles, the intestines and vital organs, the blood vessels and the nerves form the soft parts.

The blood vessels and nerves pass through all parts of the body; the fatty tissues underlie the skin and round the entrails and muscles. Of the liquid component parts the blood is the most important.

The bones of the human body, more than two hundred in number, are many of them like pipes which contain a bloody soft substance or marrow; others are flat like the outside bones of the skull, while others are spongy like those of the Vertebral Column. Each bone is surrounded by a thin soft membrane.

Many bones finish in an elastic and somewhat solid substance called gristle or Cartilage.

Naturally gristle is found in the larynx and the external ear without being attached to a bone.

As a rule every two bones are bound together by ligaments and where the two bones are intended to move one upon another the connections are called joints. Each joint is like a cover hermetically sealed and formed of masses of ligaments while the ends of the bones are lined with a transparent membrane; the joints are supplied with synovia, a greasy slimy fluid, which facilitates the rubbing against each of the ends of the bones.

Many joints, for example that of the elbow. allow of movement merely to one level whilst others like the shoulder joint, admit of movement in many directions.

The muscles which form the principal mass of the flesh are composed of bundles of filaments which affect the movements of the body and its parts. They also possess the power of shortening themselves by contraction and of returning to their former longer form by relaxation of the shortened condition.

The muscles gnerally run between two bones and at least one joint; they are attached to the bones by shoots, like strings, which are called sinews.

The result of a muscle contracting is that in becoming shorter it causes simultaneously an approximation of those portions of the body to which its extremities are attached.

Let us take for example the arm:

If the muscles of the upper arm contract while the arm is outstretched, the forearm will approach nearer the upper arm because of the bend which takes place at the elbow joint; and when the muscles relax together with the hinder ones of the upper arm, they return to their former condition.

The glands also belong to the soft parts.

Muscles of Man.

Front View.

1. Straight Abdominal Muscle. — 2. Navel. — 3. Oblique Inner Abdominal Muscle. — 4. Pectoral or Breast Muscle. — 5. Deltoid (Triangular in outline and covers the shoulder joint). — 6. Broad Neck Muscle. — 7. Head Nodding Muscle (Nutans). — 8. Biceps. — 9. Straight Femoral or Thigh Muscle. — 10. Sartorius. — 11. Two Headed Calf Muscle (Biceps of the leg). — 12. Patella (Knee Cap).

Photolithograph by J. Kösel in Kempten.

Muscles of Man.
Back View.

1. **Broad Dorsal** or **Back Muscles**. — 2. **Deltoid**. — 3. **Anguli Scapulae Muscle**. — 4. **Glutens Maximus Muscle** — 5. **Two Headed Calf Muscle** (Gastrocnemius). — 6. **Hinder Blade Muscle** (Erector Spini). — 7. **Small Gluteal Muscle**.

Photolithograph by J. Kösel in Kempten.

Their office is to separate the secretions and excretions of liquid from the blood which are partly used in the functions of the body, as, for instance, the gastric juice in the digestion, and partly leave the body, as for example the urine conducted through the kidneys, thereby getting rid of all useless material.

The glands as a rule possess one or more exits through which the rejected liquid flows. There are large glands like the liver and very small ones such as the perspiration glands which are scarcely visible to the naked eye. The secretion or excretion is either thinly liquid like urine or slimy like saliva or yet again tenacious like the wax of the ear.

There are other organs which do not secrete outwardly and yet are called glands, for instance the Lymphatic glands.

The skin which is drawn over the outer upper surface of the body is of eminent importance. It consists of the fine outer skin (Epidermis) and the deeper layer or dermis.

The outer skin (Epidermis) is covered with hairs which acquire in certain parts of the body, especially on the head a noticeable length and thickness.

We recognise the wisdom of God in the construction of man by many signs, one of which is that he has formed nails at the extremities of fingers and toes for the protection of the skin.

In the dermis or inner layer we have small hollow spaces scarcely visible to the naked eye which are called Cutaneous or Sebaceous glands; they are really small secreting tubes which discharge their fatty secretion by small ducts or pores usually into the sheath or follicle of one of the hairs. The secretion of some of these glands is a greasy substance while in others it is a watery salt substance or sweat. The skin covering the inside of the natural openings of the body such as the mouth etc. is called the mucous membrane, and one can see this transformation from skin to mucous membrane clearly in the lips and near the eyelashes. This membrane is of thin quality with a reddish appearance caused by the blood vessels showing through.

The upper surface of the mucous membrane receives a damp slippery matter by means of a secretion ejected by one of the finest possible glands. This membrane covers the upper surface of the internal passages of the body, for example the mouth, nose, larynx, oseophagus or gullet, the stomach, the intestine canal etc.

Division of the Body.

The body of man is divided into the head, the trunk, and limbs or members.

The head is subdivided into skull and face. The skull proper or case for the brain has a semi-round form and contains, in the cranial cavity, the brain.

The skull is subdivided into the forehead or frontal bone, the crown of the head, the occiput and the sphenoid on both sides.

The crown of the head (Parietal) Occiput and part of the Sphenoid are covered with the hair of the head.

The face contains eyes, nose, mouth, cheeks and chin. The ears form the division between the skull and the face.

The parts of the trunk are the neck with the back portion, the nape, the chest, the belly, the back, the loins and the pelvis with its side portions, the hips. The ridge which separates the trunk from the upper part of the thigh is called the flexor. Two large cavities filled with organs, the cavity of the chest and of the abdomen form the interior of the trunk.

Man has two sets of limbs, the arms and the legs or the upper and lower members. The head consists of the skeleton, of the face and skull which are surrounded by soft portions.

The bones of the head are tightly connected and inseparable. Only the lower jaw or mandible possesses a capacity for movement, its joints being in front of the ear and may be felt in movement.

Furthermore the nasal bones, the malar bones or cheek bones as well as both bones of the upper jaw belong to the face skeleton.

The eye sockets, the cavity of the nose and of the mouth are formed partly by the bones of the face with one another, partly by the bones of the skull as well as gristle and soft parts.

The eye sockets or orbits formed of bones which are wide open in front, go deep into the head and become narrower at the back and inwards.

At the backmost portion a small round opening gives access to the cavity of the skull through which the optic nerve runs to the brain.

The nasal duct connects the eye-orbits in the outer internal angle with the nostrils (nares).

The nasal cavity is divided into two parts by what is called the Septum formed partly of bone, partly of gristle, both halves left and right being open in front and behind.

The passage to the cavity of the mouth forms the continuation of the hinder part of the nasal cavity, while the cavity of the mouth is divided from the nasal cavity by the palate.

The anterior bony parts are named the hard palate, and the hinder movable parts the soft palate.

The bottom of the cavity of the mouth is formed of a soft substance which encloses the tongue bone (Osseous hyoid).

In the upper and lower jaw are the teeth of which the adult has sixteen top and sixteen bottom making in all thirty-two.

In each jaw are four incisors, two dog or canine teeth and ten molars.

Wisdom teeth are the hindmost molars which are cut after the 16th year of life. Every tooth consists of the visible crown and the fang; the former is chiefly of hard dental enamel; the fang is fastened in the jaw bone.

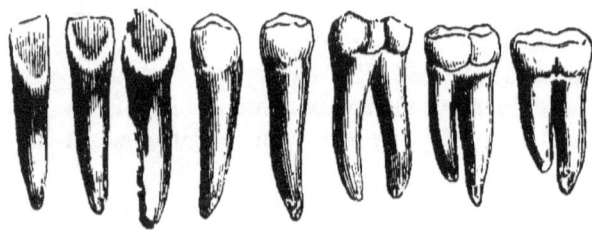

Teeth.

Dentine is the substance of the tooth beneath the crown and above the fang. In the interior of the tooth is a papilla permeated by blood vessels and their accompanying nerves.

Behind the teeth is the tongue which lies in the narrowest place in the cavity of the mouth at its hindmost part with the soft palate.

On pressing down the tongue one notices in the hinder part of the mouth the uvula hanging down from the middle of the soft palate.

The tonsils are two bluish transparent salivary glands which lie beneath the tongue; two similar glands are to be found on the lower edge of the underjaw and in front of the exterior side of this bone in the vicinity of the ear.

The fluid ejected by these glands mixes with the secretion of the mucous membrane of the mouth and forms the saliva.

The Vertebral Column or back-bone called also the Spine is the support of the trunk. It runs from the head to the pelvis and consists of a number of small bones placed one on the top of the other. Each of these small bones is termed a Vertebra and all are held together by strong ligaments.

The Vertebrae may be divided into those of **the neck**, seven in number, called **Cervical;*)** those of the back of which there are twelve and known as **dorsal;** and five large Vertebrae below the dorsal called **lumbar.** All these are known as **true Vertebrae** because they are connected by ligaments. Beside these there are others, five or six in number which form a mass called the **Sacrum** which is just below the **lumbar** and gives attachment on each side to one of the haunch or hip bones.

Each Vertebra with few exceptions consists of a rounded body or irregular ring of bone thickest in front from which several hard bony projections stand out in various directions.

The **Vertebral** Canal which contains and protects the spinal marrow (spinal cord) is formed of the bony arches of the Vertebrae lying closely one on another and in connection with the occipital space in the skull.

The ribs four and twenty in number, twelve on each side are long slender curved bones which extend from the spine and bend horizontally from behind towards the front.

The seven upper ribs on each side join the breast bone or Sternum by cartilages and are called **true ribs.** The five lower ribs do not join the breast bone and are called **false ribs.**

The breast bone or Sternum is a flat bone which extends along the front of the body downwards from the neck.

It receives the ends of the upper ribs, protecting the chest in front and sheltering the heart.

At the upper end the Clavicle is joined on which runs to the shoulder. The lower breast bone and the

---

*) From **Cervix** the neck.

cartilages mounting up to the lower ribs enclose the Cardiac region or region of the heart.

The cavity of the chest is surrounded by the pectoral casket; it is formed by the twenty-four ribs, the breast bone and the Vertebral Column.

Under the cavity of the chest lies the pit of the stomach; this is divided from the cavity by the diaphragm, a partition partly fibrous and partly muscular bordered below by the pelvis, behind by the lumbar and on the upper side by the soft parts.

The **Pelvis** is a formation of the os sacrum and both hip joints which are united in front with one another by a cartilaginous substance.

The round cavity found on the outer side of the hip bone is the joint socket which is intended for the end of the thigh bone. The cavity of the Pelvis is the lowest of the parts of the gastric cavity enclosed by the Pelvis.

The arms or upper members divide into several parts such as the shoulder, the upper arm, the lower or forearm and the hand.

The shoulder joint consists of the shoulder blade, a triangular flat bone which lies at the back of the trunk, and the collar bone in front which runs from the spiral hollow shoulder bone in this form ∿-running tolerably horizontally from the lower point of demarcation of the neck to the breast bone and finally from the upper end of the upper arm bone or arm cap.

This possesses a round semi-ball-like-surface-joint which with a joint socket lying on the exterior side of the shoulder blade forms the shoulder joint.

Among these, between the trunk and the upper arm, lies the shoulder cavity.

The bony part of the upper arm consists of a strong hollow bone, the humerus, whose lower end is recognisable by two sharply prominent sideway knobs which, with the bones of the forearm, form the elbow joint.

Skeleton of Man.

Front View.

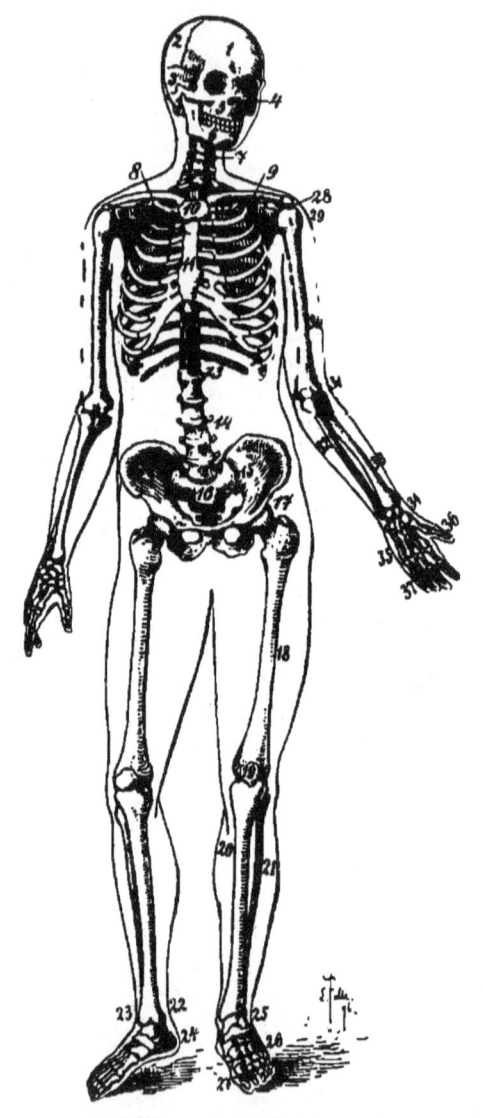

1. Frontal bone. — 2. Parietal bone. — 3. Temporal bone. — 4. Malar bone (Cheek bone). — 5. Upper Jaw bone. — 6. Lower Jaw bone. — 7. Cervical Vertebrae. — 8. and 9. Clavicle or Collar bone. — 10. Hinge of the Breast bone. — 11. Body of the Breast bone. — 12. First rib. — 13. The 12th Rib. 14. Vertebral Column. — 15. Iliac bone. — 16. Sacrum. — 17. Head of Thigh bone. — 18. Thigh bone. — 19. Patella. — 20. Tibia (Shin bone). — 21. Fibula. — 22. and 23. Inner and Outer Malleolus. — 24. Calcaneum (Os Calcis) Bone of the heel. — 25. Astragalus. — 26. Metarsus (Bone of Middle of the foot). — 27. Toe Bones. — 28. Shoulder Joint. — 29. Cap of the Upper Arm bone. — 30. Upper Arm bone. — 31. Elbow Joint. — 32. Ulna. — 33. Radius. — 34. Carpus (or Wrist). — 35. Metacarpal Joint — 36. Thumb. — 37. Finger Phaelanx.

Skeleton of Man.
Back View.

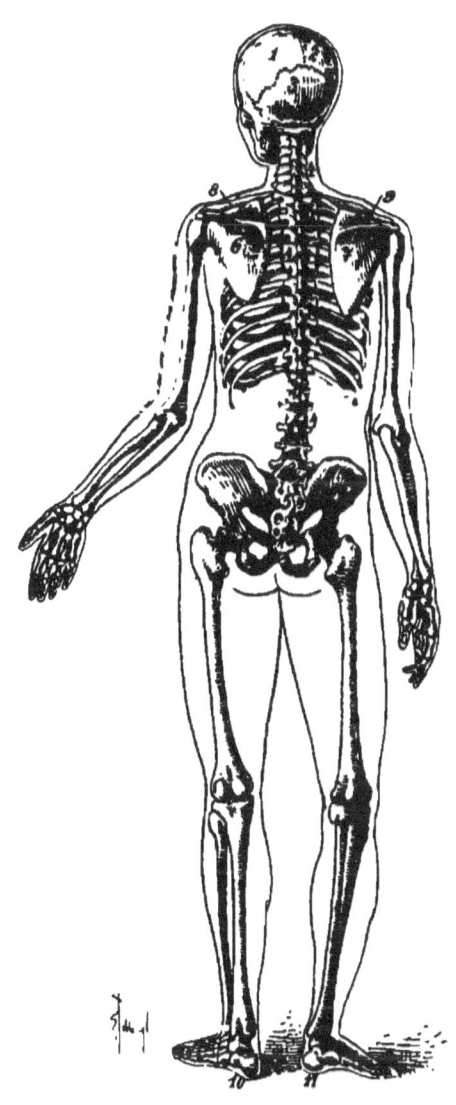

1. and 2. Parietal Bone. — 3. Occipital Bone. — 4. Vertebral Column (or spine). 5. Os sacrum. — 6. and 7. Scapulae (shoulder blades). — 8. and 9. Clavicle (or collar bone). — 10. and 11. Calcaneum (os calcis). Heel bones. For all the rest see front view of the skeleton.

Photolithograph by J. Kösel in Kempten.

The forearm skeleton consists of the **Radius** laid on the thumb side and of the **ulna** on the side of the little finger.

The upper shoulder-like extremity of the latter is noticeable at the back side of the elbow-joint outside. The radius is movable round the ulna thereby rendering possible the turning of the hand which follows its movements.

The hand consists of the carpus, or wrist, the metacarpus and the fingers or **phalanges**.

The bony part of the wrist is formed of two rows of little bones the eight carpals, which with the upper row and the lower end of the radius and ulna form the wrist joint. The **metacarpus**, the middle or solid part of the hand, is divided into the back of the hand and the palm of the hand, the latter being bordered by the ball of the thumb and the ball of the little finger.

The movement of the fingers is caused by little muscles of the hand which lie in the metacarpus and also by muscles in the forearm whose sinews extend along the carpus and metacarpus as far as the fingers.

The legs or lower members take their commencement from the hips and are divided into the thigh, or femur, the lower part of the leg and the foot.

The bony part of the thigh is the os femoris, a long strong bone, the longest and the strongest in man.

The ball-shaped-joint-extremity above, with the haunch bone socket, forms the hip joint.

The lower part of the leg with its fleshy back side, the **calf**, consists of two bones, the shin bone (or tibia) inside and the calf bone or **fibula** outside.

The **tibia** is connected at its upper end with the lower end of the **femur** at the knee joint: to this belongs also the flat knee pan (or patella) a small bone somewhat triangular yet rounded in outline. It is attached above by its broad upper margin to the tendon of the front muscle of the thigh; below, a ligament goes from its pointed lower end to the upper part of the shin bone.

The back part of the knee-joint is the popliteus, (space or hollow of the ham).

The foot joint arises from the thickening of both lower leg-bones at their lower extremity to the inner and outer bones.

The **tarsus** consists of seven bones, none of which are large, and among these are the calcaneūm and the astragalus.

The foot consists of **tarsus**, metatarsus and phalanges, the sole of the foot and the back of the foot.

When a man stands upright, the foot rests below on the balls of the big and little toe and the prop formed by the calcaneum; the last named portions are covered with hard skin.

The middle part of the sole of the foot has a thinner skin, is slightly curved upwards and is called instep. In flat feet the whole sole rests on the ground in standing, because in these the instep has sunk so far.

The sinew which runs like a rope from the muscles of the calf to the hinder end of the calcaneum is called the Achilles sinew.

Entrails are the soft portions enclosed in the large bodily cavities of **the trunk.** The heart and the lungs belong to the entrails of the breast.

The lungs are composed of lobes lying over one another, three on the right and two on the left called right and left lung and contain numberless little bladders or air cells.

From these issue delicate tubes which combine into larger tubes and run finally into the bronchi from which each one leads into the five lobes of the lungs.

The trachea is a combination of two further tubes one of which receives the bronchi of the right and the other the bronchi of the left.

Lungs of Man.
Front View.

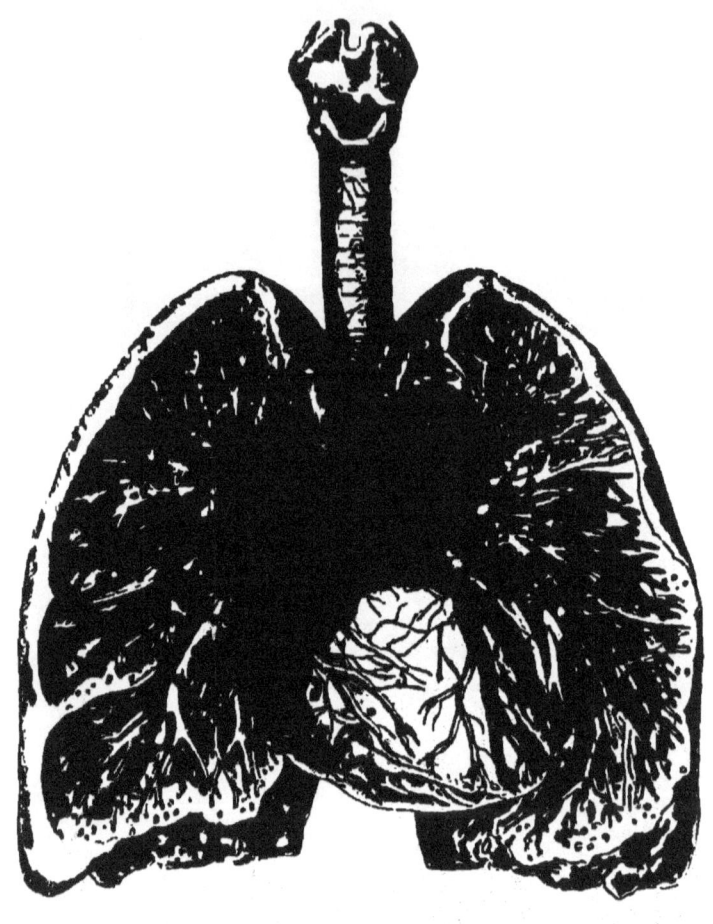

1. Larynx. — 2. Right. 3. Left top of the lungs. — 4. Trachea. —
5. Heart as it lies in front the lungs. — 6. Aorta. — 7. Pulmonary artery
with the ramifications coloured blue. — 8 Vena Cava. — 9. and 10. Base
of the lungs.. The pulmonary veins are coloured red; the bronchi for
contrast are yellow.

From an original drawing by the Artist Eug. Felle after an illustration in the work "Der
Mensch" by Dr. John Ranke, Bibl. Institute in Leipsic.

Photolithograph by J. Kösel in Kempten.

The trachea runs in the centre of the neck and goes or passes into the larynx which opens below the mouth and constitutes inspiration and expiration by means of expansion of the chest.

The lungs are covered on their upper surface by the pulmonary mucous membrane, a fine skin, the inner wall of the cavity of the chest, but is covered by the chest skin or pleura.

Respiration is the unbroken activity of the lungs which conducts to the body the air necessary for life. It brings the oxygen of fresh air into contact with the blood as it passes through the lungs.

Inspiration occurs by the drawing in of air into the trachea and its branches which causes the air cells to expand like air bubbles.

In **expiration** the used up air is ejected from the little air cells and the expanded lungs again contract.

The inspiration and expiration can be seen outwardly by the movement of the chest in its expansion and elevation, in its depression and contraction.

The air taken into the lungs contains plenty of fresh oxygen while that discharged contains much less (as 16 to 21) and is foul with carbonic acid, nitrogen, and vapor.

This is plainly to be seen in cold weather when in breathing out, the moisture thickens into vapour.

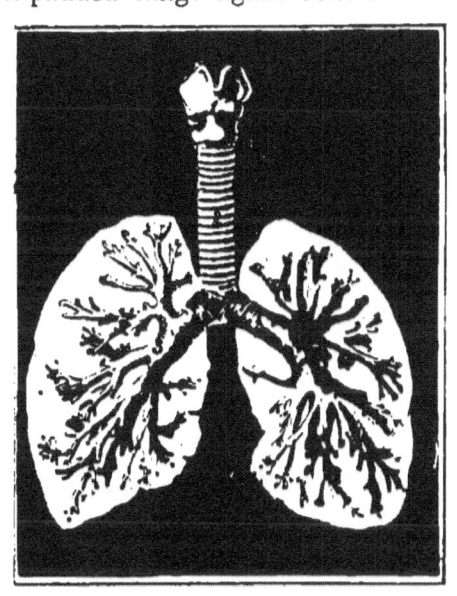

The lung tree.

1) Larynx, 2) Trachea, 3) and 4) Ramifications of the Trachea.

It is clear also that where many people are congregated the air is perceptibly corrupted.

9*

Respiration or the act of breathing is performed about sixteen times a minute; but during rapid movement such as brisk walking, running, mountaineering and the like and also in some diseases this increases.

Healthy children breathe quicker when at rest.

The Larynx and the Voice.

The voice is produced by expiration in the larynx which last with its gristly formation perceptible at the neck, encloses the vocal cords or bands of elastic tissue which run close together from front to back. In relaxed condition however during rest they lie so far apart that the air breathed in can pass freely through them; they can be brought closer together by means of little muscles in the larynx. By the air breathed out, and which streams past, they are set swinging, and according to the tension of the cords high or deep tones are produced.

With the help of the organs lying in the mouth, for example, the tongue, lips, palate and teeth the voice is impelled to speech.

The Nobler Organs of the Human Body.

The Heart and the Circulation of the Blood.

By inspiration part of the air, in the interior of the lungs, mixes with the blood which flows uninterruptedly through the human body during its whole life and is called the circulation of the blood.

The blood is liquid, red and contains the colourless blood-water and many minute corpuscles.

Many of them have a coin-like appearance and are red, while others are ball-like and colourless.

For this reason they are distinguished as red and white blood-corpuscles.

Blood Vessels of Man.

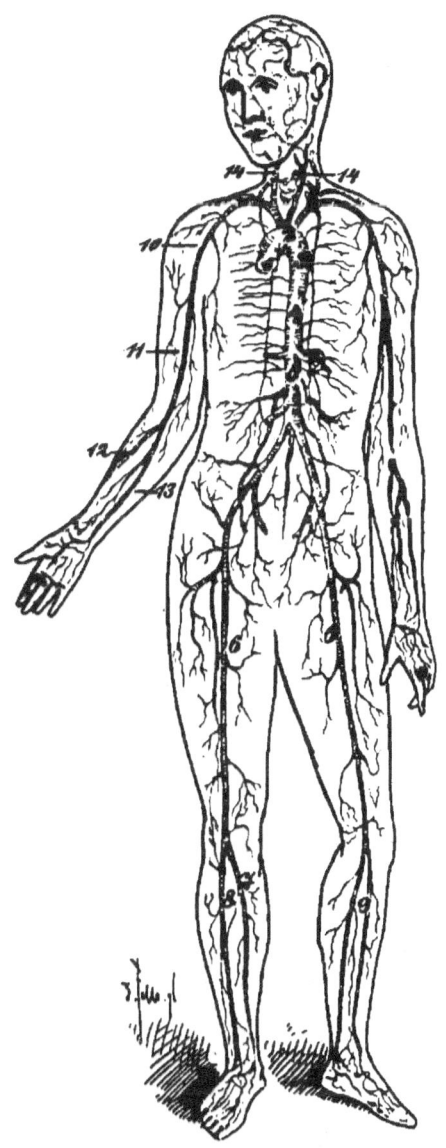

1. Ascending aorta or artery. — 2. Arch of Aorta. — 3. Descending aorta. — 4. Aorta. — 5. Abdominal Aorta. — 6. Femoral. — 7. Posterior tibial. — 8. Anterior tibial. — 9. Fibula artery. — 10. Axillary. — 11. Brachial. — 12. Radial. — 13. Ulna. — 14. Carotid.

Photolithograph by J. Kösel in Kempten.

The blood is partly in the heart, partly in the veins which are tube-like, elastic blood vessels.

On leaving the heart the blood passes through arteries, capillaries and veins. By the first the blood runs out of the heart into the body; by the latter the blood returns from the body back into the heart.

The heart, surrounded by the pericardium, a membraneous formation in the form of a bag, lies in front in the left half of the cavity of the chest between the ribs, the breast bone and the wall of the breast.

It is about as large as a fist and ball like. Part of the heart lies on the wall of the chest, the other is covered by the lungs.

The heart is formed by masses of muscles and encloses a hollow cavity which is divided by one partition lengthways and by another across into four parts.

Both the upper ones

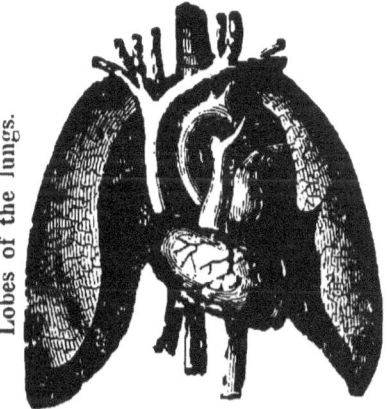

Lobes of the lungs.

The heart.

are called **auricles** because they look outwardly something like dog's ears; the lower are called **ventricles**, the walls of the first are thin, while those of the latter are very strong in order to perform their functions.

The auricles collect the blood while the ventricles act like force pumps: the right one pumps the blood through the lungs, but the left forces it through the entire body.

The big artery or aorta issues from the left ventricle from whence the course of blood starts on its nourishing circulation upwards, after which it curves backwards to the vertebral column and from there downwards to the pelvis where it divides into two arteries for the two lower members. (Iliac arteries).

At the spot where the aorta forms an arch near the heart branch off the arteries of the head, neck and arms, and downwards the arteries of the entrails of the breast and stomach.

Various arteries bifurcate into the minutest of vessels called capillaries which spread over the body like a net.

When the capillaries unite then we got the little veins; from these arise the larger ones and from these the very big Venae Cavae of which the one conducts the blood from the head, neck and arms, while the other conducts the blood from the remaining parts of the body back to the right auricle of the heart.

The circulation between the right heart auricle and left auricle is the greater or body circulation.

In the lesser circulation the blood leaves the right side of the heart by an artery and passes through the capillaries of the lungs returning through veins to the left side of the heart.

The circulation system comprises the conveyance of the blood to and from all parts of the body by means of **arteries** which take blood from the heart; by **veins** which take blood to the heart; and by **capillaries** which convey the blood to the tissues and intervene between and connect the ends of the arteries and veins.

The **capillaries** are distinguished from the other two vessels by the absence of muscular fibre in their walls.

The circulation of the blood is caused by the contraction of the heart which forces the blood into the arteries and causes the phenomenon which is known as the **pulse.**

In adults this contraction takes place about seventy times a minute; in old people not quite so often; in children more frequently.

This contraction alternates between the ventricles and auricles of the heart.

The Heart of Man.

Left side View.

Front View.

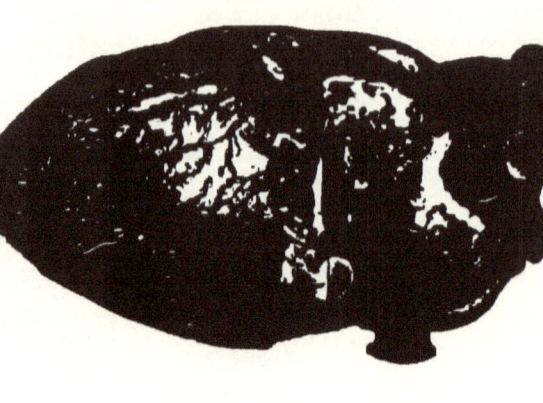

1. Aorta. — 2. Pulmonary artery. — 3. Left Auricle. — 4. Vena Cava Superior. — 5. Vena Cava Inferior. — 6, 7, 8, 9. Pulmonary Veins. — 10. Left Ventricle. — 11. Right Ventricle. — 12, 13. Blood vessels for the nourishment of the heart.

1. Large artery or aorta. — 2, 3, 4. Principal branches of the arch of the aorta. — 5. Left auricle. — 6. Right auricle. — 7. Right ventricle. — 8. Left Ventricle. — 9. Left ventricle. — 10. Right ventricle. — 11. Pulmonary artery. — 12. Vena cava superior. — 13 and 14. Blood vessels by which the heart is nourished.

The part coloured yellow is the fat which surrounds the heart.

From an original drawing by the Artist Esq. Feile after an illustration in the work "Der Mensch" by Dr. John Ranke. Bibl. Institut in Leipzig. Photolithograph by J. Kösel in Kempten.

First, the auricles contract driving the blood into the ventricles, then the ventricles contract driving the blood into the arteries; then there is a pause during which the auricles are filled by the blood poured in from the veins. The contractions are then repeated.

When the auricles contract, the valves open letting the blood into the Ventricles. Then when the ventricles contract the valves close, preventing the blood returning to the auricles; and the other valves open letting the blood out into the arteries.

As soon as the ventricles cease contracting, these last valves are closed by the elasticity of the arteries which, having been over distended by the contraction of the Ventricles, contract, attempting to drive the blood back.

The closing of the valves makes the return of the blood from the bodily and pulmonary arteries impossible. The effect of the valves is to allow the blood to move in one direction.

Some diseases so change the form of the valves that they can no longer shut up: they are called Valvular diseases. As a rule they lead to a disturbance in the circulation of the blood, because, on the expansion of the ventricles, the blood streams partly back and causes obstruction.

One consequence of the expansion of the arteries is that the pulse rises; this may be tested by pressing a finger on the wrist.

Diseases, especially fever, or agitation, influence the strength and number of the pulse beats as well as alter the colour of the blood.

The oxygen inhaled from the air is partly absorbed by the capillaries of the pulmonary air-cells in the blood and combines with the red corpuscles to make the blood scarlet. This colour continues while the blood passes through the left auricle, the left ventricle and the arteries in the body: the oxigen, however, is distributed by the capillaries of the greater circulation in the surrounding tissues and in its place a similar quantity of carbonic acid is imbided causing the blood to be black-red.

Thus coloured, the blood flows through the veins, the right ventricle, auricle and pulmonary veins, exchanging in the lungs the imbibed carbonic acid for new oxygen.

By a process similar to combustion carbonic acid is generated in the corporeal tissues and ejected by expiration or breathing out.

In addition to veins and arteries we have in our bodies certain vessels known as lymphatic vessels which contain a colourless liquid which has exuded from the capillaries. These vessels exist in all parts of the body: they have proper walls but they originate in mere channels left as it were between the other tissues; and thus whatever is cast loose must find its way into the lymphatic vessels.

The centre of the lymphatic system is a vertical canal which ascends in front of the vertebral column and is called the **thoracic duct.**

Into this duct all the lymphatics ultimately empty themselves except those of the right arm and right side of the head which empty themselves into a small vessel called the right lymphatic duct.

The lymphatic glands are rounded structures and consist of finely divided lymphatic vessels. They are scattered in various parts of the body and vary in size from that of a lentil to that of a bean: they contain innumerable little cells like those of the blood corpuscles.

Between these cells the impure matter accompanying the lymph is filtered.

If matter is introduced which has been absorbed through wounds or diseased tissues, inflammation of the lymphatic glands sets in.

The Nervous System.

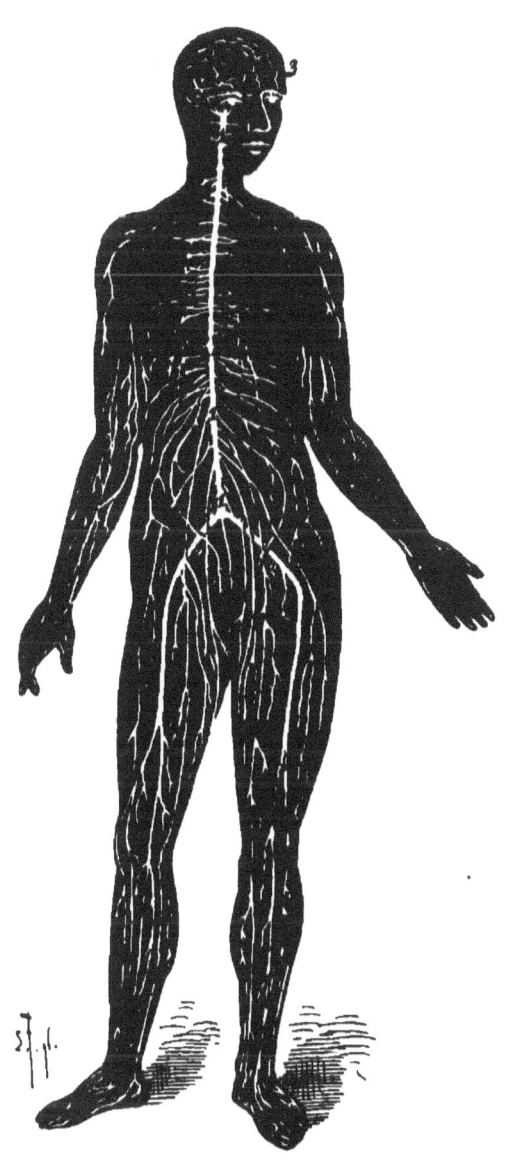

1. Spinal cord. — 2. Haunch nerves. — 3. The brain.
The illustration shows the distribution of the nerves.

Photolithograph by J. Küsel in Kempten.

In the cavity of the abdomen are the digestive organs, the renal organs and the spleen.

To the digestive organs belong the stomach, the alimentary canal and the pancreas.

The Stomach.

The stomach is a somewhat long or pear-shaped membraneous bag so curved that its upper surface is concave. It is one of the alimentary organs below the diaphragm, the others being as we have said the liver, the intestines and the pancreas.

The gullet or oseophagus is a nearly straight membraneous, muscular tube leading from the pharynx*) to the stomach. It opens near the wide left end, or fundus of the stomach, which part is called its **cardiac** part and forms a link between the mouth and stomach.

The right half of the stomach contracts itself in the form of a funnel up to its continuation to the gut.

The part of the stomach leading into the intestine is the opposite end to the cardiac and is called the **pylorus**. Here the stomach is shut out by a sort of valve or ring-like muscle.

The Stomach,

1) Mouth of the stomach, 2) The exit of the stomach, 3) The stomach, 4) Pit of the stomach.

The Alimentary or Intestinal Canal

is a sheath formed of membraneous walls certainly six times as long as the body of man. It is divided into

---

*) Pharynx is the swallow.

the narrow small intestine and into the wide great gut. The former begins by the stomach and is there called the duodenum and then with many windings fills the larger part of the cavity of the abdomen. To the right in the lower abdomen in the vicinity of the hip bone it winds round into the great gut; the commencement of it which lies under the soft covering of the abdomen forms a bag like protuberance downwards to the so called blind gut or caecum.

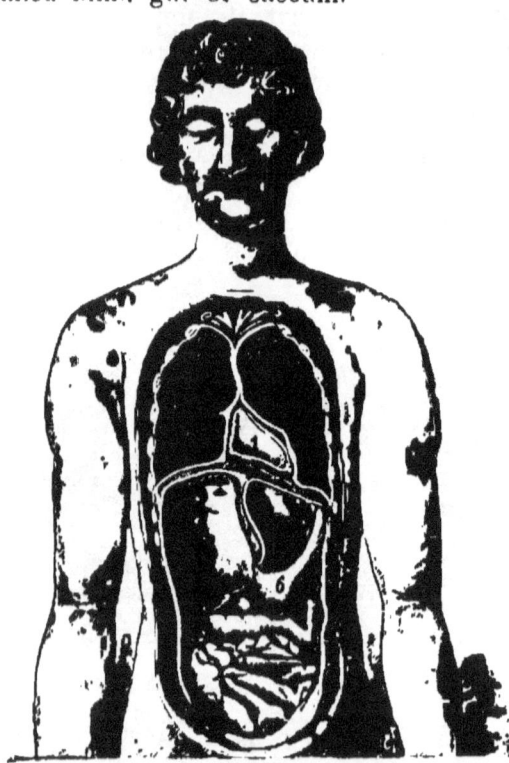

Front view of the viscera of the chest and abdomen.
1) Heart, 2) Left lung, 3) Right lung, 4) The liver, 5) Spleen, 6) Stomach, 7) Intestines.

This last has attached to it, near its end, a very narrow hollow process called the **vermiform appendix** which is frequently rendered dangerously inflamed by indigestion.

From the caecum the great gut ascends in the abdominal cavity, passes transversely and then descends to terminate in the rectum. In its transverse course it passes very close to that part of the small intestine called the **duodenum.**

All the alimentary organs below the diaphragm are invested with a delicate serous membrane which also lines the inner wall of the whole abdominal cavity: it forms a very large sac, folded in a complex manner and contains a serous fluid. This sac by which the viscera are attached (as in a sling) to the front wall of the vertebral column is called the **peritoneum.**

In the cavity of the abdomen and formed by the peritoneum there are many membranes and ligaments which enclose and fasten the viscera; these are called the **mesentery.**

In front of the cavity of the abdomen behind the wall of the stomach is the so called **caul** which forms a membraneous apron in front of the viscera. In corpulent people this is greatly encumbered with fat.

The Liver

lies to the right of the stomach under the diaphragm in the upper part of the cavity of the abdomen within the cartilages of the ribs. It is a large brownish red organ smooth and convex above, concave and uneven below. It is divisible into certain parts or lobes which are defined partly by grooves and notches, partly by ligaments and blood vessels.

The liver secretes a yellowish or brownish fluid which is stored in the gall-bladder which has a fine exit in the duodenum; this liquid is called the bile. In the same place where the bile enters the intestine, there also the pancreas sends its juice into the intestines.

The Digestive Organs

subdivide themselves first into the canal which begins at the mouth, runs through the cavities of the body and ends at the anus, and secondly in the glands, whose secretions debouch into the digestive canal.

Whatever is eaten by man is digested through this canal.

Digestion is therefore the- extraction and absorption of the nourishing matter of the food by the blood necessary for the maintenance and growth of the human body.

The useless material in the nourishment passes as excrement through the anus openings of the body.

The nourishing materials consist of albumen, stuffs containing sugar or starch. and fatty matter.

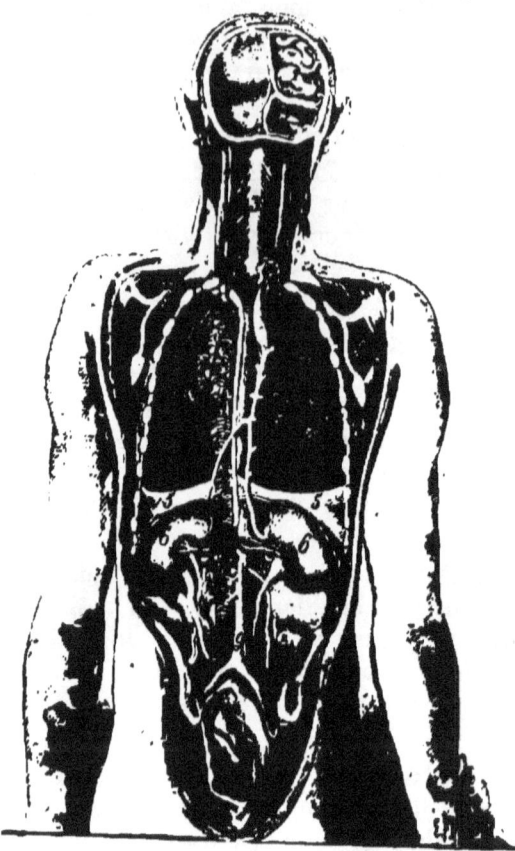

Back view of the opened head as well as the viscera of the chest and abdomen.

1) Large brain, Cerebrum, 2) Lesser brain, Cerebellum, 3) Pharynx, 4) Lungs, 5) Diaphragm, 6) Kidnoys, 7) Aorta, 8) Stomach, Renal vessels, 9) Vena cava.

The digestion of the latter is managed by the bile, that of the albumen by the secretion of the stomach, and that of the starch and sugar by the saliva of the mouth. The dissolution of the food begins in the mouth where it is cut up and ground by the teeth. By decreasing the size of the food digestion is greatly aided. By the movement of the tongue the palate and the gullet the food is swallowed down the oseophagus and through this reaches the stomach: the larynx ascends and becomes covered by the epiglottis which prevents the entrance of the food into the larynx and wind pipe.

When the nourishment has reached the stomach, the gastric juice is secreted and mixed with the contents of the stomach; during this operation the pylorus is shut and the food can find no entrance into the intestine. The process of digestion lasts according to the kind and substance of the food from one to six hours.

Now when the food has undergone the process of digestion in the stomach, it passes in a pappy form through the pylorus into the intestine.

By the addition of the bile, the pancreatic juice and the gastric juice, this pap becomes almost liquid and is coloured brown by the bile. This liquid now makes its way, with the help of the snake-like curves of the intestines through the small intestine, it gradually acquires solidity in this process especially in the great gut and finally leaves the body, as solid excrement, by way of the rectum.

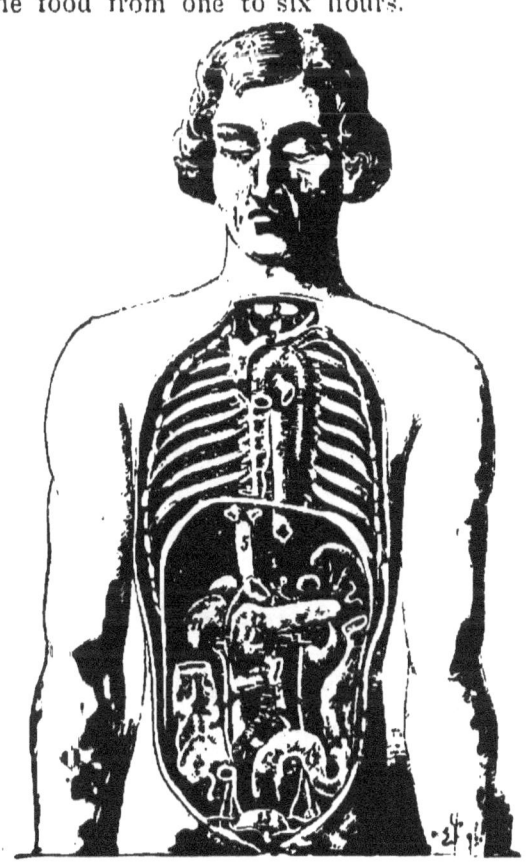

Front view of the viscera of the chest and abdomen after the removal of some of them,

1) Ascending aorta, 2) Arch of aorta, 3) Descending aorta, 4) Oesophagus, 5) Lower vena cava, 6) Thyroid gland, 7)—9) Upper vena cava with ramifications, 10) Spleen, 11) Kidneys, 12) Pancreas, 13) Caecum, 14) Great gut, 15) Bladder.

This thickening has its origin in the absorption of the liquid parts into the lymphatics and blood vessels: in the small intestine especially, the so called chyle or nutritive juice is conducted by the lymphatic vessel of the mucous membrane into the blood and through that to the cellular tissue which absorbs it.

In this way the old tissues are maintained and new ones formed as well.

The activity of the tissues depends on a constant use of the chemical stuffs of the body.

A kind of "burning" process known in chemistry as oxidation goes on in the spaces between the capillaries, and it is to these that the fluids of the blood part with the pure and nourishing particles received from the food and take up from them the impure and used up par-

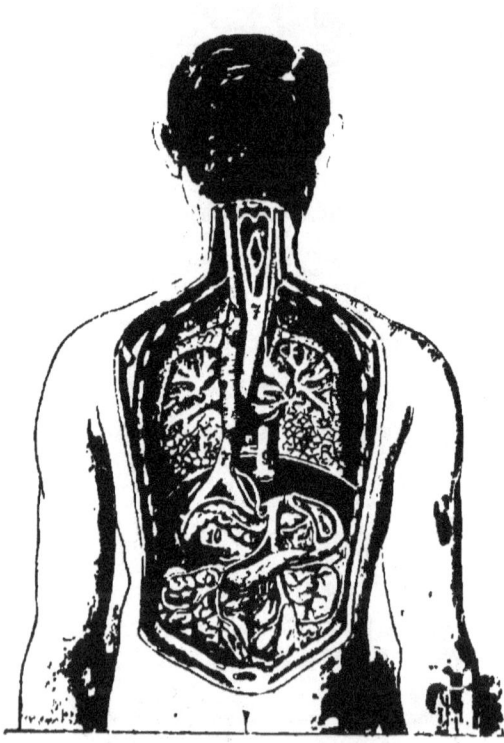

The viscera seen from behind.

1) to 4) Lungs, 5) to 6) Bronchi, 7) Upper end of the stomach, 8) Aorta, 9) Vena cava, 10) Stomach, 11) Intestines.

ticles which have to be removed from the body.

In this way both strength and warmth are maintained in the body: it is necessary for the maintenance of life to administer fresh nutritive matter and to form new cellular tissues, for as they grow old and useless they are burned.

The cause then of animal heat is the burning that goes on between the network of capillaries throughout the body, the muscles and glands being the main sources of heat. To make it clear; the power and heat of the human body are formed by the burning of tissues, built up out of food, with the oxygen got from the air.

The temperature of a healthy person averages 98.4⁰ F.

Illness may cause it to rise; exhaustion to fall.

Diminution of warmth is increased by the exudation of sweat. In the heat of summer the body cannot cool itself sufficiently in the hot air: therefore the pores of the skin give out more sweat than usual.

The clothing of the body protects it in colder air from too great a cooling and from chills. Strong physical exercise causes the temperature to rise.

On the approach of death the body grows cold in consequence of the cessation of activity on the part of the cellular tissues.

During the activity of these tissues certain materials pass into the blood while non nitrogenous food passes off in watery vapour or carbonic acid gas by breathing or by sweat and urine.

The Water or Urine.

is a clear fluid variously coloured according to its contents and mostly bright. Out of the body in the air it gets dull and thickens into corruption.

With invalids various ingredients pass with the urine which often give explanations of the illness and its properties.

The ejection of the urine occurs by the two kidneys which are bean-shaped, brownish-red glands lying near the lumbar column on the posterior wall of the cavity of the abdomen embedded in fat. Out of each kidney runs the Ureter which opens into the urinary bladder. This lies in the pelvis in front of the rectum and conducts the urine out through the urethra.

The Spleen.

The spleen is one of the organs which lie beneath the diaphragm at the cardiac end of the stomach: it is of reddish-blue colour, of firm substance and assists in the formation of blood. In illness it often increases in circumference and size, so that one can feel it externally.

The Brain and Nervous System.

Beside the regular functions of the body to which those things already described belong and which take place regularly without depending on will, there are yet other manifestations of life in the human body assuming that the man is conscious; these are known as perceptive powers. The principal organs of the nervous system are the brain and the spinal marrow.

The brain fills up the whole cranial cavity: it is surrounded by little skins and is a soft tissue permeated by many fine blood vessels.

The mass of the brain is divided into the upper narrow grey crust and the white kernel which latter has grey portions in the centre and many hollow places filled with liquid.

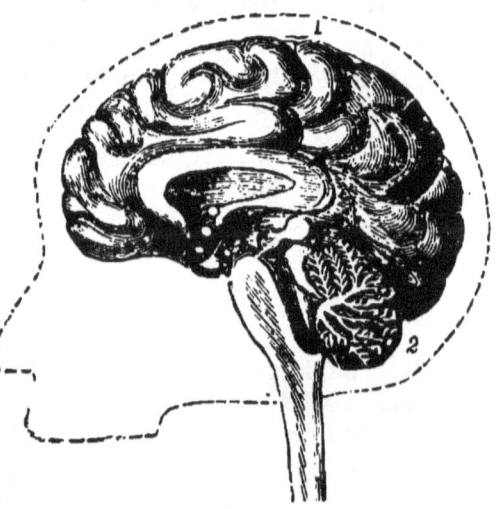

The brain (in section).
1) Cerebrum (great brain), 2) Cerebellum (lesser brain).

One subdivides it into the big brain and little brain; both are divided into a right and left half by a lengthways cut.

Again it is divided into lobes called, after the situation they occupy, forehead lobe, temple lobe, middlehead lobe, back of the head lobe, spirals.

Between the spirals and lobes many little curved furrows run up to the brain: they are not so deep as the lengthways and straight furrows.

The spinal marrow has its seat in the Vertebral column: it is a soft tissue like the brain and of cylindrical form, white on the upper surface and gray internally.

At its upper end the lengthened spinal marrow passes into the brain in the cavity of the skull; the hollow spaces of the brain continue also in the spinal marrow canal which runs through the entire organism from top to bottom.

In the brain and spinal marrow there are innumerable ganglionic cells from which issue the tender nerve filaments which unite in white bundles, nerve channels. These last form the white of the brain and spinal marrow mass. They frequently cross themselves in the brain and then run into the spinal marrow as strings close together in the form of a bundle.

Out of the brain and spinal marrow are formed the nerves, white strings as thick as needles, which by many divisions and ramifications run through the various parts of the body in fine, scarcely visible, threads and lines.

The ganglionic cells of our brain and spinal marrow are the seat of consciousness, of our perception, of our actions and sensation.

The nerves are the agents between the gangliac cells and the different parts of the body; they receive impressions and carry out the actions permitted by the will.

If individual parts of the brain are injured externally or internally, it may result in the loss of one of the capacities of feeling or movement, for example, an injury to the part called lobe of the brain may cause loss of speech or the severing of the nerve of sight and so on.

Of the individual nerves the noteworthy ones are the twelve pairs of brain-nerves which go direct to the brain and the nerves of smelling, seeing, hearing and tasting as well as the nerves of movement like the eye muscle, face and tongue nerves.

Out of the spinal marrow thirty pairs of spinal cord-nerves issue of which each has a posterior and an anterior root.

The anterior root is composed of those nerve threads which run from the brain and spinal cord to the organs of movement; on the other hand, by the posterior root are formed those nerves which conduct the sense of perception to the brain and spinal cord.

The result of disturbances in the anterior root is paralysis of the affected muscles while the same in the posterior root result in injury to the department of sensation.

The Senses and their Instruments.

Man has five senses : sight, hearing, taste, smell and touch. Each of these has its distinct instrument whose office it is to conduct the external impression and its sensation to the nerves of the brain.

The Eye and the Sense of Sight.

The organs of the sense of sight are the eyes. The eye consists of the apple and its aiding and protecting mechanism. The apple of the eye lies in the cavity of the eye enclosed in soft tissues and is of the size of a large cherry: it is connected with the brain by the nerve of sight or optic nerve which comes out of the skull through a hole into the cavity of the orbit and reaches to the posterior wall of the apple of the eye where the nerve filaments dissolve: on each apple of the eye is a thick covering and inside it a jelly-like substance, the Vitreous humour.

The covering divides itself into three layers; the external is a firm, china-white, hard skin and forms the protective cover to the inner part of the eye; a portion of the front surface is visible in the so called white of the eye.

The middle layer is the choroid. This is of delicate tissue, black internally, and within it the little blood vessels ramify which run into the apple of the eye.

The inner layer, the **retina**, is a delicate, tender membrane of the filaments of the optic nerve.

In front on the hard skin is a circular glassy section, the cornea, through which the light penetrates to the interior of the eye.

The part of the choroid lying behind does not join the cornea, but is spread between the anterior chamber and the interior of the eye like a curtain in the space caused by its curving.

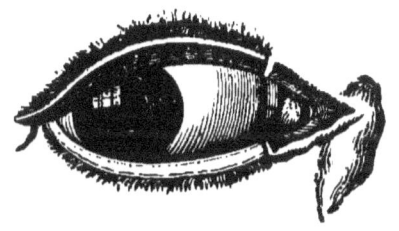

The Eye.

This part of the eye is the iris skin so called on account of its various colours; hence the eyes may be blue, brown, black or gray.

In the midst of the iris skin is the seeing hole or **pupil**, the "black spot of the eye".

The pupil contracts or expands and lets, as it were, more or less light fall into the interior; it is therefore a contrivance by which too strong a light is avoided.

Immediately behind the pupil is the **crystalline lens** formed of a glassy compact tissue which is like a magnifying glass in front and curved behind.

10*

Through the crystalline lens the rays of light, falling through the cornea and pupil on to the back ground of the eye, unite in a picture which the retina reproduces exactly.

In an ordinary lens-arch only the rays of light falling in the eye on the back ground are united; on the other hand, those rays running to the eye one after the other do not meet until they get beyond the background of the eye. The rays coming from a longer distance meet parallel in the eye. The lens is so constructed that with the help of the optic muscles its curve can be strengthened so as to be able to unite with the rays coming from nearer parts in the background of the eye.

With long sighted people the eyes have such a narrow lengthways diameter that the lens must increase its curve for the union of the parallel rays in the retina, but rays running separately are not in a position to unite on the back ground of the eye, therefore the picture on the retina is confused and such eyes can only recognise objects a long way off. This infirmity can be remedied by convex spectacles.

In short sighted persons the eyes are so formed that the parallel rays unite themselves in front of the back ground of the eye, therefore can only produce clearly pictures of objects close at hand: because the rays of light coming from close at hand strike the eye divergently and therefore can be united by the lens further off than the parallel rays.

In such short-sighted eyes the optic power can be increased by concave spectacles which divide the rays of light before they reach the eye. If the lens of the eye is clouded gray by injury, illness or age, and the optic power is diminished or lost, cataract ensues.

Optic capacity can be restored by operation if the lens that has become opaque is removed; yet the eyes must be provided with properly polished spectacles with convex glasses.

By means of the optic muscle one can turn the apple of the eye in its socket in various directions and see many objects at the same time; by turning the head one can over-look a yet larger space.

If both eyes turn at the same time on an object lying close at hand, they survey it from several sides and therefore more easily perceive its shape.

Any disturbance in the action of the optic muscles causes the condition known as squinting. To protect the eyes from external injuries certain contrivances are provided, as for example, the eye-lids which protect the ball of the eye from the entrance of foreign bodies such as dirt, dust, splinters and such like; the eyebrows also serve to protect the eyes.

The inner surface of the eye-lids is covered with a mucous membrane reflected over the eyeball and called the **Conjunctiva.**

For the removal of objects which occasionally gain entrance to the eyes the **Lachrymal** or tear-secreting gland is provided: the tears flowing into the membraneous bag from the tear ducts make their way down a canal which traverses the lachrymal bone into the **nares** or nasal cavity.

In crying the lachrymal excretion is very great and mostly external, as well as in certain inflammations of the eyes.

The Hearing.

The hearing lies in the ear. By its means man perceives **sound**; each ear is divided into the **external** which receives the sound; the **middle** which conducts it further and the **inner** which distinguishes the sound.

The external ear consists of the concha and the outer hearing canal which leads inwards into the skull where it is kept flexible by the aural wax excreted by delicate glands.

A delicate, elastic skin, the membrane of the tympanum, separates the external ear from the middle one; the latter consists of the cavity of the tympanum, the Eustachian tube and the little bones of the ear. The cavity of the tympanum (or tympanic, so called because of its connection with the drum of the ear) is a small hollow space filled with air and is connected with the nose by a delicate pipe called the Eustachian tube.

The little bones of the ear, called from their shapes, hammer, incus or anvil, and stapes or stirrup, are connected with each other by fine delicate little strings. The **inner ear** consists of the three archways, the Vestibule and the Cochlea.

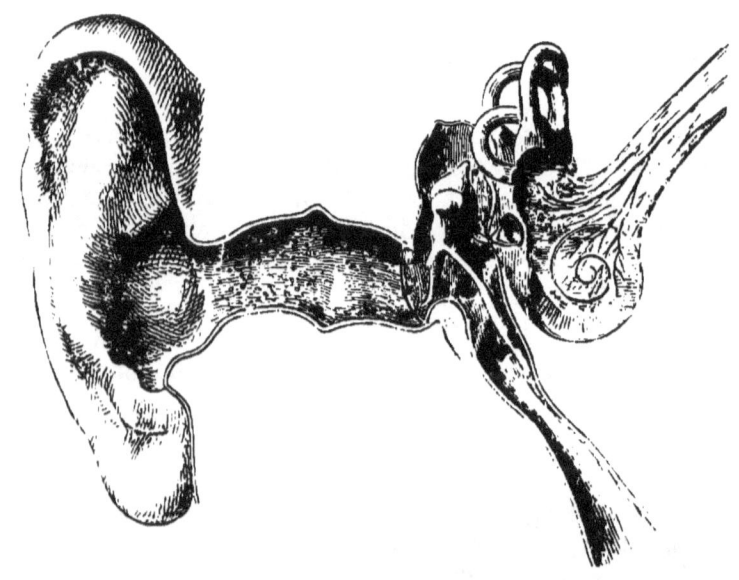

The Ear.

Within the cochlea is the end of the nerve of hearing which runs from the brain to the ear, divided into many tiny threads lying close together.

By the sound which penetrates through the concha and the acoustic canal into the middle ear, the membrane of the tympanum is set swinging; this motion, communicated through the little bones of the ear, sets the liquid of the ear in motion, affects the nerve filaments in the inner ear and conducts the sound to the brain.

In order to protect the membrane of the tympanum which is liable to split at too harsh a sound (a cannon shot for example) we recommend the opening of the mouth which will give the air a wider passage to the Eustachian tube.

The Sense of Taste

lies in the tongue and is called forth by materials which are soluble in the saliva; the tasting nerves form the connection between the tongue and the brain.

The Sense of Smell

has its seat in the smelling nerves which run from the brain to the walls of the cavity of the nose and ramify in the mucous membrane of the nose.

The sense of smell enables us to perceive volatile things which penetrate to the nasal mucous membrane by means of the air.

The Sense of Touch

has its seat in the nerve of feeling which lies in the skin. By this nerve one experiences heat, cold, pain, pressure and the like; by it therefore we distinguish sensation of temperature, feeling of pain, touch and pressure.

The brain, wearied often by the many experiences of the senses, requires rest which it gains by sleep. During slumber. breathing, circulation of the blood and digestion go on uninterruptedly; on the other hand consciousness and the voluntary muscles suspend their activity. The respirations are deeper and fewer during sleep. The duration of sleep is regulated according to age and activity. Little children, weakly persons and old people need more sleep than others; healthy adults can manage with from six to eight hours of sleep.

Care of the Human System.

General.

he human body is certainly the most beautiful and complete machine of all the things created by God; no single part in or about it but has its high signification, and even if but a single part is lacking the same thing happens as in a machine, it either works with great difficulty or is quite unable to work at all.

If man would be healthy, enjoy long life and fulfil the duties imposed on him by God he must look after the whole body as well as the single parts of it with the greatest care.

Unfortunately there are thousands and thousands of struggling men and women who attach no importance to their bodies and have no knowledge how to care for them.

It is of frequent occurrence that people's eyes are opened only when it is too late; when an infirmity has gained a footing which might easily have been avoided. This is why I again give my contemporaries and posterity a few hints and counsel as to how they should care for their general system in a proper manner as well as for the individual parts of the same.

The Eye.

Of all the members of the body the eye certainly takes the first place; it is the greatest blessing possessed by man.

Who is able to fathom what it is to be blind?

Yet it is the few and not the many who take proper care of their eyes. Innumerable are the people one meets who are suffering with their eyes!

A proper and natural care of the eyes should be practised from childhood.

Happy are the children who are born and brought up in the country or on the mountains! They have the best medicine for their bodies and eyes. The fresh health-giving pure air provided here by Mother Nature, together with the lovely view of the green meadows, forms a varitable cordial for the eyes.

Under these conditions they are strengthened and braced and are exercised for long distances as well as near.

If we take the contrary and pass on to the children dwelling in big towns, in large or small streets, in damp and unhealthy dwellings, what a difference we find! As a rule the whole body is crippled and weakened and the eyes do not escape injury.

Certainly they are not developed or strengthened and have no chance of gaining good sight for distances. If light and air are taken from the eyes no one can expect them to be healthy.

For the eyes above all it is necessary they should have good fresh air and light in abundance, for above all members of the body they are the hardest worked.

In caring for the body the eyes should receive special attention that they may become strong and capable of resistance.

As the half bath renders good service to the system so does the eye bath given four times a week benefit the eyes: the manner of applying it is described in "My Will" page 48.

These baths are very strengthening and cleansing and no one should grudge the trouble. Those who have weak eyes cannot be too strongly urged to take eye baths. They are excellent and very effective for the recovery and sustaining the seeing power and for the cleansing of the eyes; the baths may be composed of different herbs, as for instance, wormwood, fennel, eyebright; and their preparation and use is explained fully in the book already mentioned. It is worth notice that compresses of these herbs are of the utmost service.

For bad eyes it is extraordinarily good to administer one drop of pure natural honey.

The numerous examples of eye diseases that I have seen here in Wörishofen induce me to specially recommend the care of the eyes.

Many have come here with weak eyes and many who were close upon blindness, yet by the application of these simple remedies their trouble has greatly diminished and many returned home completely cured. A great cause of mischief is working in the twilight or with a badly burning petroleum lamp! This foolish habit obtains specially among students and school children.

Nothing can be more harmful than this foolish practice, many diseases of the eye have been traced to it.

It must be clear to every one that working in the dusk weakens the eyes and sensibly diminishes the power of sight; yet now-a-days the over pressure of school work or professional business forces people to work by night and many, to save light, work at dusk forgetting at the time that by so doing they ruin their eyes.

I implore parents never to permit their children to read or write in the dusk. I equally oppose immoderate work by a light because it is harmful to the eyes.

Men should rest at night so that on the next day they may return to their duties with renewed strength.

A question frequently asked is "should one wear spectacles?"

My answer is that if great care has been taken of the eyes from one's youth this question is, as a rule, superfluous.

If the eyes are weak and one thinks it absolutely necessary to use eye glasses, at least put it off as long as possible for when once taken to it will be very difficult do to without them.

I never like to see young people wearing spectacles, for I wonder where it is likely to lead in later times; of course the glasses must from time to time be changed for stronger ones.

For old people spectacles are often indispensable because, in their case, the eyes are much weakened.

Never omit taking an eye bath every morning when washing as prescribed in my book, for in so doing the eyes are strengthened and protected from many evils.

The Care of the Ears.

How many people there are who are partially deaf or altogether so! In searching for the cause one finds that it may be traced to imprudence, uncleanness or a bad illness. The question is, how can one take care of one's ears? First of all, secure cleanliness by washing or syringing the ears with cold water or a decoction of pewter or shave grass (Exquisetum Arvense. L.)

Further one should never voluntarily set oneself in a draught for that has often led to great and severe mischief. Anyone susceptible in this respect should stop the ears lightly with wadding.

With the ears as with other parts of the body bracing should be exercised.

This is the way the ears should be cared for; it is cheap and simple and if practised daily, there will be no complaint of partial deafness.

The Care of the Nose.

Seeing that the nose is not made of wood but on the contrary is often very susceptible and is the cause of various diseases, it requires careful attention.

It is all the same whether one has an abnormally large nose or only a tiny one, they must receive equal care and attention.

It might be supposed that snuff-takers bestow the greatest care on their noses seeing that they handle them constantly: scarcely a minute passes when out comes the snuff box and pocket-handkerchief.

This however is by no means taking proper care of the nose.

If snuffling is necessary a decoction of shave-grass or cold water is the best.

A very nasty habit practised by children and adults alike is a constant picking of the nose by the fingers: I warn such people that many diseases, among which are cancer and lupus, are brought about solely by this practise. Children indulging in this bad habit should have their hands rapped so that they will not continue it.

For the Care of the Hair

read the part entitled Hair and Nails.

The Care of the Teeth.

The teeth require special care quite as much as any other part of the human organization. Only those who have suffered tooth-ache know the value of good teeth. Who could have thought that one single bad tooth had the power to cause so much pain or render the night sleepless.

If therefore the teeth are capable of causing such intense pain it must be doubly necessary to bestow care upon them. Here again we must compare the past with the present.

How many people one sees now-a-days who, scarcely grown up, have their mouths full of gaps or false teeth: what is more harmful to the teeth than the many dainties on which so large a number of mothers, in their blindness, rear their children! What can be more damaging to the teeth than the modern fashionable food! Such food, was quite unknown in former days; people then lived reasonably, simply and moderately and but few were afflicted with bad teeth.

I once went to a cemetery in which a grave had just been opened wherein lay a man of from fifty to sixty years of age. As the head came in sight I saw to my great astonishment that all the teeth except two were in existence.

Our teeth are so absolutely necessary to us for masticating our food and assisting the digestion that it behoves mothers to see that they are not spoilt by fashionable food. Ask an old man how difficult he finds the mastication of food and how happy he considers himself if he still possesses a couple of stumps!

Sugar plums, sweet meats, sharp and pungent spices, vinegar, pickles and the like are poison to the teeth, therefore at all costs to be avoided or at least only used in quite small quantities.

Many children have bad teeth when quite young solely in consequence of eating too many sweets.

One of the chief points to be observed in the care of teeth is cleanliness. Every decent man washes himself at least once a day and it would be very little extra trouble to cleanse his teeth at the same time.

Every morning they should be cleaned with a simple little brush and cold water or shave grass tea, and to keep them in really good condition they should receive the same attention in the evening before going to bed.

To the use of tooth powder I am not wholly averse but I would advise the greatest care in the choice. See that it is made up of harmless materials which will do the teeth no harm.

Should any of my dear readers not care to give out money for tooth powder let them take a fresh piece of charcoal, crush it to powder and use it in the same way as other tooth powder. It would answer every purpose.

The warning not to take things too hot has frequently been given and I wish to emphasize this most emphatically, for not only does it injure the teeth but also the organs of the neck and stomach. Further avoid all sharp pungent foods and beverages, for by the use of these the teeth may be completely ruined.

Many stupid people during a fit of tooth-ache fly at once to acids and tinctures to allay the pain, or pick the gums with a knife or sharp instrument. If a toothpick is necessary let it be of soft wood or bone.

The cracking of nuts is also very bad for the teeth: if nutcrackers are not at hand use hands or feet rather than the teeth.

In this place I wish to refer to the chapter concerning tooth-ache in the first part of "My Will", and as a completion add to it yet this, that camomiles soaked and laid on the cheeks in a little bag renders excellent service in tooth-ache.

The Care of the Face.

The care of the face is zealously practised in many circles, but in what manner?

The feminine sex specially take the greatest trouble to acquire a beautiful complexion and spare neither paint nor powder in the process. It is sickening to approach such a fashionable goose.

The face is best cared for by daily washing in quite cold water.

If the chlorotic creatures would take a simple nourishing diet such as whole meal bread, Brenn-soup etc., instead of adulterated food, the colour of the face would be much improved.

The Care of the Mouth.

The care of the mouth is really included in the chapter on the care of the teeth, yet I wish to add something more.

There are, as one knows, many people afflicted with a very evil smelling breath or as they say in the vernacular they smell from the throat.

This is not only very unpleasant for the victims themselves but still more for those who approach them.

People are not always to blame for this, for not unfrequently the unpleasant odour arises from a disordered stomach or it may be caused by the remains of food left behind in the cavity of the mouth, which last shows negligence in the care of the mouth or teeth.

Such people should on no account omit to rinse their mouths constantly with cold water, or shave grass and sage tea.

They owe it to themselves and to those among whom they live. I advise those in whom the bad smell arises from the stomach to use the "Juniper berry cure" or to drink wormwood tea every other day and to avoid such food as is likely to produce a disordered stomach.

The Care of the Throat.

As every individual member of the whole human body must receive special care and attention, so also the organs of the neck and throat.

Our experience of the many and often serious diseases of the throat teaches us how compulsory the care of the neck and throat is.

The manner of accomplishing this is very simple — the chief thing being extreme cleanliness.

Gargling the throat every morning with cold water or shave grass or sage tea should never be omitted; it not only strengthens the organs but also renders the throat capable of resisting growths etc.

Always, if possible go with the neck uncovered so that the air may penetrate everywhere and never use comforters or shawls as unfortunately so many country people do in the present time. They do this thinking that the wraps will guard the throat from chills whereas they do just the opposite; for by this careful wrapping up, the organs of the throat are yet more weakened and should the smallest current of air penetrate to them, some unpleasantness will be the result. Those, on the contrary, who understand how to brace the throat will avoid colds and chills and should, by chance, either appear will know how to get rid of them speedily.

The Care of the Organs of Speech.

Certainly all people may consider themselves fortunate who possess clear ringing voices with which they may speak, sing, and lecture clearly and harmoniously.

Unfortunately few only are so blessed and it is one of the annoyances I am subject to that people coming to me with reports, especially students, speak in such a manner that it is with difficulty I can understand.

Among a hundred people there are scarcely three who really open their mouths and have learnt to speak properly.

Either they speak as fast as a running machine wheel or in a disagreeable manner squeeze the words out, neither of which is right.

All people should accustom themselves to speak slowly, distinctly and intelligibly. With such it is a pleasure to exchange words.

"What is the best way of practising the voice?" I am frequently asked.

Go into the country, especially in the neighbourhood of woods! There you will hear the children in their light-heartedness singing and shouting with such joy that it is a pleasure to listen to them.

Seeing that this practise strengthens the voice I have advised many to go out into the forest and there wake the echo. At the same time I would observe that these exercises must not be overdone and that they must be undertaken gradually.

I have constantly made these exercises in my youth and I believe I have to thank them for my powerful voice; without praising myself I may say that as regards strength and endurance of voice few can surpass me. In spite of my seventy-six years I am still able to give several lectures daily without experiencing the least inconvenience.

I only wish all, especially those who are destined for public careers, would exercise and strengthen their voices in this way. Such exercises help not only the voice, but aid in developing the body, if they are practised in combination with water applications; half baths and constant wading in water will not fail in their effect.

Tea of green fir-cones or tincture of the same is not to be forgotten. I recommend it highly to preachers and professors. It would be well to take a cup full of the tea every three days — the taste is not good but one soon gets accustomed to it. Of one thing I am sure that a cup full of this tea is worth more than a quart of beer.

I would advise all whose professions demand much use of the voice, whether it be on the boards or elsewhere, to take a few drops of pimpinella tincture before speaking.

Further, guard against debilitation!

The more the neck is exposed to the air the stronger will be the organs of the throat: therefore never wrap it up with shawls and comforters.

The Nervous System.

The nervous system is a very complicated arrangement in the human body, and one that calls forth our wonder at the omnipotence of our God.

The nerves, with the brain and spinal cord for their centre, go out into all parts of the body and effect sensation, thought, will, feeling; in short all that goes on in the human body. From the nerves is derived the will power of man which in nervous invalids is naturally much weakened.

Unfortunately the nervous system does not escape disease of various kinds; that known as the "disease of

11*

the period" viz., **nervousness** or hysteria is brought about mainly by the fashionable way of living.

To fight energetically against the destruction of the nervous system, nothing more is necessary than a proper bracing, simplicity of food and an avoidance of the indulgence of passions common to men.

~~~~~~~~~

The Care of the Lungs.

These are very sensitive organst and require the most careful attention.

Every thing that could possibly injure them should be avoided, for people with diseased lungs are miserable.

To those who lead a sedentary life I would say a few words. In working as in walking try as much as possible to maintain an upright position, for nothing is worse for the lungs than a bent curved one, which prevents the lobes of the lungs from properly expanding. It is necessary for these people to expand their chests frequently during work and it is indispensible that they get frequent exercise in the open air. Where do we find more people with diseased lungs than in towns?

It is a very dangerous habit to drink quickly when in a heated condition and it is one which has claimed many victims.

Therefore I recommend to readers of this chapter great prudence, and advise them earnestly to keep the body free of weakness by a suitable bracing: I know no better way of caring for it than by bracing and I wish that every one would practise it and then there would be a chance of mitigating human suffering.

As it is with the members of the body already described, so it is with those remaining, bracing is now and always the principal thing to maintain health and strength. This being so, it is not necessary for me to enter more into detail as to the care of the remaining organs of the body except to say that bracing extends both to food and clothing as well as to the body itself.

I trust that these few rules of conduct may enter into the hearts of my readers: may each and all be prompted to the care of the soul as well as to the proper care of the whole bodily organism, for when we neglect the latter we sin no less against God.

Fifth Part.

A few Plants from God's Garden.

O, all ye green things upon the Earth, bless ye the Lord!

Carline Thistle. (Carlina Acaulis L.)

his thistle bears different names according to the locality in which it grows. The soil on which it flourishes is mostly rough, calcareous or quite unfruitful.

The blossom is of a beautiful silver white, and about as large as a five shilling piece; its stem, which is spinous, is a foot high. The root is frequently used by veterinary surgeons for diseases of cattle, but it has a good effect also in many human infirmities. Country people place it among the magical herbs, side by side with St. John's Wort.

In all the experiments I have made with this plant, its effect has been to strengthen and purify internally as well as externally.

Any person suffering from indigestion or weight on the stomach will obtain help at once by taking a tea made of this root; for the sap clears all hurtful stuff out of the intestines, cleanses the whole body, especially the kidneys, so that the urine passes out clean.

As the Carline thistle dissolves and disperses all watery corrupt matter it is an effectual remedy in dropsy and in whitloes.

The Carline thistle is dried and ground to powder; take as much of this as will lie on the point of a knife twice a day, or make a cup of tea of the powder using as much as will lie on the point of two knives; take one third of the tea three times a day.

A tincture may be made from the Carline root by drying it, cutting it up small and leaving it during five days in good brandy or spirit.

The Carline thistle. (Carlina Acaulis L.)
a) The whole plant, a fifth of its natural size,
) and c) Blossom, natural size, d) Fruit,
e) Fruit cut across, slightly enlarged.

Take of this tincture from twelve to fifteen drops in a spoonful of water twice a day. The Carline root is also a good remedy for external use in eruptions; it purifies the skin and causes the scars, if there be any, to vanish.

For this purpose the Carline root is cooked in two parts water and one part vinegar, or better still, in one half water and the other half wine. With this decoction the diseased part is washed frequently.

Chickweed.   (Stellaria Media L.)

This herb is considered by country folk, among the noxious herbs or weeds.

It grows on road sides and waste places; it is strong and covers whole tracts of land.

As a household remedy this herb is often of great service and may be used both internally and externally.

Chickweed may not inaptly be called a lung herb
seeing that it acts absorbingly, conducts matter out and
renders good service in vomiting of blood and discharge
of blood by coughing, as well as in hemorrhoids and
congestion of the kidneys and bladder.

A tea is prepared from this herb and a small cup
of it taken twice daily, or still better divide the cup
into five or six portions.

In congestion of the lungs the effect of this tea is
still more purgative if a third or fourth part of wine is
added.

The addition of pewter or shave grass or lanceolate
plantain is good and advisable.

The best results, however, are obtained by mixing
the sap of chickweed with honey and taking a tea spoon
full six or eight times a day.

The juice of chickweed is easily obtained by bruising the fresh herbs, which are very juicy, in a mortar;
this being done put the mass into a muslin bag and
press the juice out.

A large quantity of chickweed is not necessary to
obtain a moderate amount of juice. The juice must now
be cooked with honey over a gentle fire keeping it well
stirred so that the steam mounts up.

Such a preparation of chickweed-juice and honey
will last a long time.

Chickweed is very effective used externally for open
injuries, eruptions and old corrupt sores; for this purpose, boil the herb, dip a cloth in the decoction and lay
or bind it on the sore which will, by it, be made clean
and whole.

Small eruptions should be washed daily with this
decoction: at first they will look rather worse than better
for the application, but this is a good sign for as soon
as the eruption subsides the cure begins.

It is also the same with sores; the outflow will, at first, be stronger but afterwards healing begins.

One of the most pitiable of diseases is certainly **Lupus**, yet one gains excellent results by the use of this decoction of chickweed especially if an addition be made to it of pewter grass, ribwort and wormwood.

Chickweed tea is also a good remedy in troubles with the cornea because it cleanses the eye and increases the brightness of it if washed three or four times daily.

Burr. (Lappa Officinalis L.)

This is one of the plants classed as a weed. It grows by the wayside and in waste places looking very much like a decayed journeyman. Playful children work much mischief with it by sticking the burrs in each other's hair or woollen dresses. However much this plant may be despised as a weed, it is worthy of all honour as a curative herb and may be used in many ways. From the leaves a tea is prepared which is an excellent remedy for gastric ulcers. I recommend it also to those who suffer from gastric inflammation and from bad digestion. Three spoonsful of this tea morning and evening is a sufficient dose. The leaves are large, ovate and somewhat pointed at the top; the stem of the leaf is rather thick and somewhat rough. The tea is very cleansing and cooling and carries off gas. It should be pure, that is, without sugar or other ingredients.

The leaves may be dried and preserved in boxes; they may be cut up before drying or powdered **after**; the stalks of the leaves may be used with them but not the stems of the plant. This tea is good for gargling and rinsing the mouth; if there are blisters in the throat or the lips are sore, they will be quickly healed by this tea which has excellent healing power for all sores.

For external sores use compresses of the tea, and if at the same time the whole body is operated on by water, it will quickly throw off all impure matter.

A salve may also be prepared from the leaves of the burr by reducing them to a fine powder and cooking it in lard for a short time. It is better still if the

Burr. (Lappa officinalis L.)
a) Flowering stem, b) Root, one fourth the natural size, c) Spiked corona leaves, d) Hollow, hermaphrodite blossom, e) Seed-corn, enlarged.

sap is pressed out of fresh leaves and mixed with lard: this is done by first cleaning the leaves then chopping or crushing them in a mortar, the juice is then easily obtained by pressing the bruised mass in a muslin cloth between two boards.

The effect of such a salve is most cleansing and healing.

The root of the burr is rather thick and woody and contains a little sap: it may also be boiled down for tea and has an equally healing effect on sores and it is especially curative for bad heads in children.

Our forefathers used the root of the burr to promote the growth of the hair. I myself have made many experiments in this direction and have come to the conclusion that for those whose hair is falling out and where the roots of the hair are not dead, the root of the burr will afford great help.

Rub the skin of the head four or five times a week with the decoction of burr root or with oil prepared from it, the latter is difficult to obtain quite pure, the decoction, however, penetrates into the pores, makes the skin healthy and in this way promotes the growth of the hair, just as plants prosper if one makes the upper surface of the ground spongy.

The seed is even better for tea than the plant itself and is excellent for sick headache brought about by indigestion or gas which may have risen to the head.

Although in a healthy condition we can dispense with such a remedy yet it should be a matter of deep thankfulness that a plant possessing such power of healing and cleansing should be so abundant and within every one's reach.

Centaury. (Erythraea Centaurium L.)

There are two sorts of Centaury, the lesser and the greater; it is with the former we have to deal. It grows mostly on dry spots, on sandy, chalky soil, in the margins of fields and meadows and in the openings of woods, and is from ten to eighteen inches high.

On the stalk grow several branches springing upwards and bearing terminal bunches of flowers of a
beautiful violet-red colour whilst the slightly fleshy leaves
are lancet shaped.

The plant flowers from the end of June to the end
of August. It is best to gather it when flowering because at that time it has most sap.

Frequently, merely the flower with the upper twigs
are gathered, but more often the whole plant is cut and
dried in the shade and preserved in a dry place during
the winter. It has, like wormwood a bitter taste.

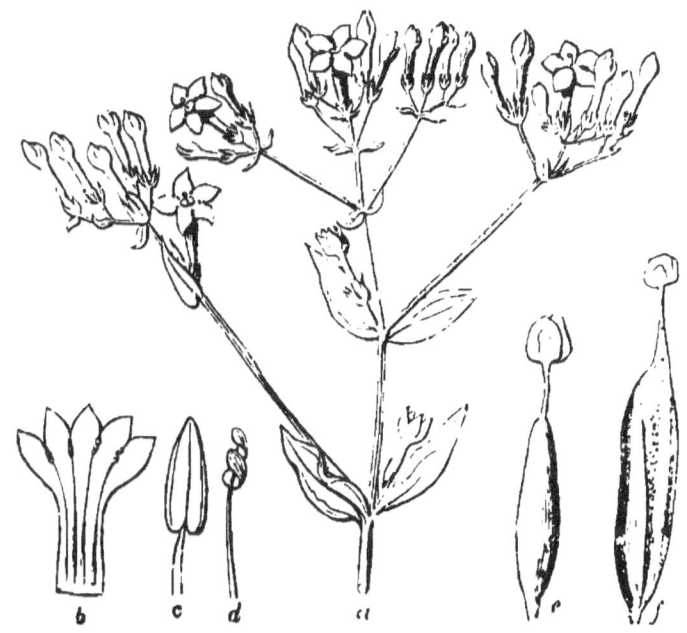

Centaury herb. (Erythraca Centaurium L.)

a) Flowering stem. b) Cut flower pressed flat, c) Young pollen vessel, d) Old pollen
vessel, e) Pistil, f) Ripe fruit.

Many centuries ago Centaury was treasured as a
curative plant. In my youth housewives used to gather
it regularly and preserve it; I remember that it was
used in many cases with most successful results; for

example, if a child had fever in consequence of a catarrh or cold the mother at once made a tea of Centaury of which a small cup daily had to be drunk. The people had the most complete faith in this remedy and to say the truth it very rarely failed.

Anyone complaining of stomach-ache, of indigestion, of loss of appetite, of heartburn, of headache, took a small portion of centaury tea four or five times a day and in a very short time the evil was removed.

As a remedy for purifying the blood the centaury plant was held in high favour among housewives and Soldiers.

Those suffering from jaundice or from liver and kidney troubles took centaury tea with a little wormwood in it as the most effectual remedy for their complaints, and they were seldom disappointed in the result.

For an ordinary cough one took, as a rule, a tea of wormwood and sage; if, however, whooping cough set in then a tea of wormwood and centaury was taken with the best results.

It is of common occurence to find both children and adults suffering from whitloes.

In such cases tea made of centaury and wormwood was taken and the whitloes were quickly banished.

For small sores, eruptions, impurities of the skin and pustular formations the general practise was to drink a cleansing tea of centaury herb and sage and the result nearly always justified the expectations.

I myself have tested the effect of this dwarf or lesser Centaury herb in all the diseases here specified and in others as well, and I have come to the conclusion that our forefathers were perfectly right in their appreciation of the plant which fully deserves its name "thousand gulden herb".

The greater Centaury herb also has its own special effect in absorbing and extracting diseased matter. It is used therefore principally in eruptions, herpes and similar diseases where a purifying remedy is desirable.

I have found it very effective in congestion of the lungs and kidneys, in vomiting of blood and expectoration of blood.

My prescription for these is a small cupful daily of this curative herb-tea, a spoonful every two hours.

The great Centaury herb grows by the road-side and is considered by country people as a good vegetable.

It is sometimes as tall as half a yard and has a violet blue flower.

Still more effective than the tea is the tincture of which one takes from fifteen to twenty drops on sugar in a spoonful of water, two or three times a day. The decoction of this herb can be used with good effect on open wounds.

~~~~~~~

Knot Grass, (Polygonum Aviculare L.)

Frequently that which in ordinary life is little or not at all regarded is often of much value: thus it is with knot grass.

It grows by the roadside, in meadows and even in the streets, so that often in walking it is trampled under foot, for which reason Germans call it **Wegtritt**. Its properties are to absorb, extract and cleanse, and for this reason the plant may be used in many ways with the greatest success.

For those who on slight provocation vomit blood it is of the utmost benefit; it should be cooked in wine and taken in small quantities, viz. a spoonful every hour: should spitting of blood set in steadily a cupful may be taken.

Knot grass is equally good as a remedy for gastric ulceration and gastric bleeding in which cases it is taken as tea in small portions; for the bleeding, a small cupful of this tea is quite sufficient.

Knot-grass-tea acts well upon the liver and kidneys, but best of all, however, in bladder troubles and stone in the bladder.

The bladder is purified by it and the stones are more quickly banished by it than by any other remedy.

I once made a report on this herb and especially noticed how, by its means, stones in the bladder could be removed.

One of the audience, a Hungarian gentleman, had for years suffered intensely with stones; at the close of the lecture he at once went to the next field and tore up a bundle of knot grass which he took to his house-keeper, begging her to make a potful of tea with it; that same evening he drank a cup of it and repeated it next morning and evening, and even on the second day he brought to me quite a collection of stones big and little, enough to fill a tea spoon, which had been extracted by the use of this tea, and his sufferings were already less. Eventually he was completely cured by its use.

Knot grass.
(Polygonum Aviculare L.)

At the same lecture a second gentleman was present who also had been a great sufferer from stone in the bladder; he also used the tea with like result; so in fact did twelve patients then staying in Wörishofen, and all were unanimous in the opinion that there was no better remedy for stone trouble than tea of Knot grass.

The effect may be still increased if, in preparing the tea, half red wine and half water be used.

Since I delivered my lecture this herb has been used for similar diseases and mostly with good effect.

Pewter or shave grass operates strongly on the stones, but Knot grass is more gentle in its action and therefore it can be used for a longer time in the most harmless manner. Not only is knot grass taken internally of great benefit, but it may be used externally with the best results for sores, fistulas, swellings and even for lupus.

Make compresses of the knot grass decoction or wash the affected parts with it.

If knot grass alone has such a good effect, it is infinitely more powerful when in conjunction with other herbs; thus we can combine broom with knot grass for an effective remedy in gravel or stone troubles, and wormwood and juniper berries with knot grass for liver troubles.

Frequently knot grass is called the little knot to distinguish it from the Great Knot.*)

The great knot was from earliest times regarded as a very effective household remedy for open wounds, especially for ulcerated feet and legs.

Bruise the plant in a mortar, and lay it on the wound. By this means the heat is quickly lessened and healing soon sets in.

---

*) The young leaves and stalk of this plant are very juicy; when however they are older the leaves become hard and dry and the plant is therefore called in the vernacular "Old Woman"

Pewter or Shave Grass. (Equisetum Arvense L.)

Pewter grass is a weed somewhat acid and not held in very high estimation among certain people: house-wives use it mostly for cleaning glass and household vessels and husbandmen look upon it with disfavour because it cumbers their land especially if the soil be poor.

On the other hand as a curative agent in the home it is of great value. It may be used in many ways; for its properties are to absorb, to cleanse, to draw together and to strengthen.

Pewter-grass-tea acts very favourably in vomiting of blood.

A short time since a man was seized with severe bleeding of the lungs and vomited quite a pint of blood. It streamed out of his mouth, his whole face was blue, the pulse almost ceased to beat, and for some time there seemed no chance of saving his life.

Then I quickly made a cup of pewter-grass and tormentilla tea and when the vomiting of blood ceased for a minute I gave him some spoonsful of it and the bleeding ceased: and when after three or four hours the bleeding began again, it was checked in the same way as before. Thus for several days this tea was con-stantly used till at length the bleeding altogether ceased.

I knew of no better remedy in the emergency than this tea: yet I advise caution in taking it, for too much produces bad results as the following example will show.

An invalid had gastric bleeding; the Doctor ordered him to drink this tea but in much larger portions than I should have prescribed.

The bleeding certainly was quieted by the tea, but in a short time he was in so much pain that he could neither eat nor drink and the tea drinking had to be discontinued.

In this instance it was clear that too much had been taken, one cupful in vomiting of blood and one

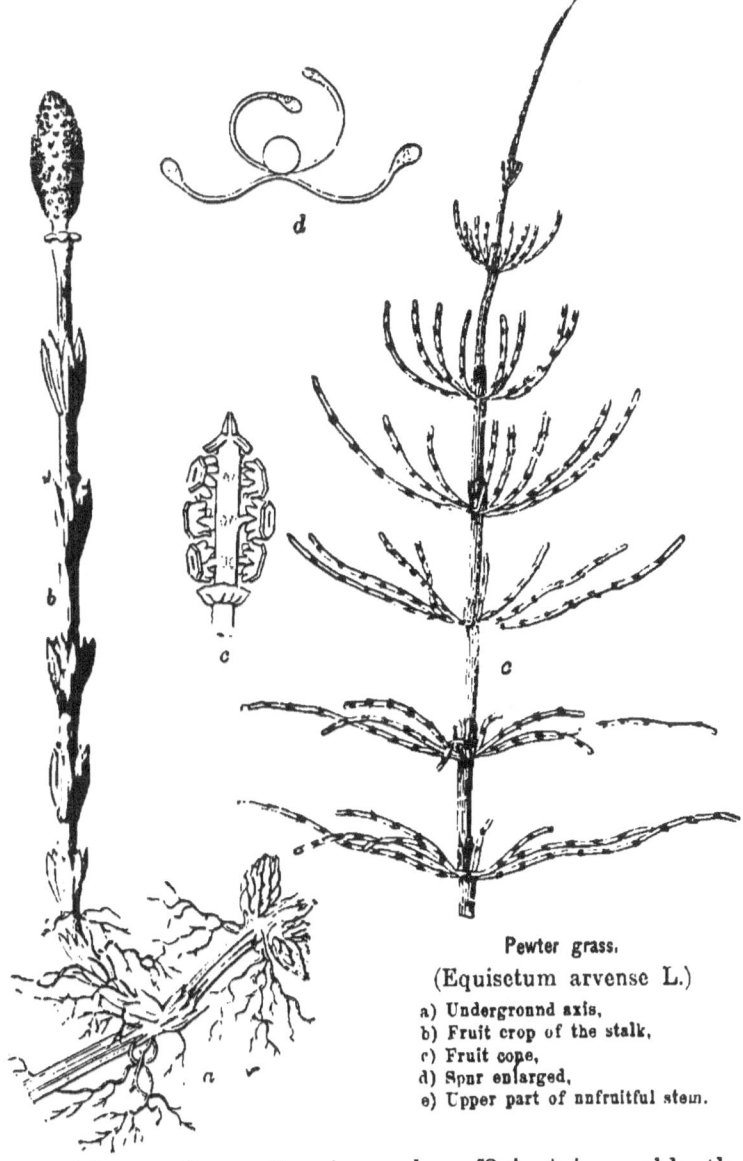

Pewter grass.

(Equisetum arvense L.)

a) Underground axis,
b) Fruit crop of the stalk,
c) Fruit cone,
d) Spur enlarged,
e) Upper part of unfruitful stem.

spoonful every hour after is amply sufficient to enable the wound to cicatrise and heal.

I was once summoned in the middle of the night
to a lad of seventeen who was bleeding to death. I
found everything in the greatest confusion and the bed
and the room covered with blood, while the patient from
weakness was neither able to speak nor even make a
sign.

I ordered a cup of pewter grass tea to be made as
quickly as possible and gave it him to drink; this at
once stopped the bleeding and as a precaution I ordered
him a spoonful of the tea four times a day: the lad
became well and strong and is still so after ten years,
and has never had a return of the bleeding.

Pewter-grass has good results in kidney and bladder
troubles, for example, an old man of seventy years of
age got a chill and, in consequence, retention of urine
set in. He shrieked with pain and I advised him to
take a close-stool vapour bath; a handful of pewter
grass was thrown into hot water and poured into the
close-stool; upon this the patient sat and in about
twenty minutes two pints of water passed.

Pewter-grass has the same good effect in all internal trouble.

If a patient suffers from retention of urine, I advise
him first of all to take a cup of the tea and each hour
or even half hour after, to take a spoonful. As a rule
this answers, but if the result be not wholly satisfactory,
place a two-fold cloth which has been dipped in pewter
grass water on the region of the bladder when the effect
will be quick and sure.

As pewter-grass used internally is unquestionably
curative, so it is most successful when used externally:
for example, foul spreading sores can certainly be cleansed
by pewter-grass and at the same time, by its property
of drawing together, be cured.

This is seen especially in Lupus, in Herpes generally, and in similar diseases of the skin. In such afflic-
tions, compresses of pewter-grass decoction should be
laid on and renewed every two or three hours.

Instead of the compresses, frequent washing may he employed. In this way the sores are healed by degrees and the intolerable itching and irritation, occasioned by the eruption, very much diminished.

People, who from time to time have eruptions on the face, would do well to wash it two or three times with pewter-grass water; this would both cleanse and heal the skin.

Pewter-grass water is an excellent remedy for new and old wounds and it is surprising how quickly the former heal when washed out with this decoction.

The results are similar if it be used for tumours, gumboils, inflammation and other injuries to the gums, palate or throat.

There is scarcely any gargle superior to pewter-grass-water, for it cleanses from slime and both cools and heals the injured parts.

Again those suffering from polypi will not find a more effective remedy for extracting and absorbing the obstructions in the blood; indeed their formation and further development may, by this decoction, be prevented.

So it is clearly seen that pewter-grass is by itself an excellent remedy for many and varied diseases and infirmities; but when used in combination with other herbs its excellency is greatly increased. It is much to be regretted that this plant is, in general, so little known and valued.

The Onion. (Allium Cepa L.)

I know three kinds of onions, first a priest called Onion whom I have cured, and another man of the same name: as the one is zealous in the care of souls, so is the other in husbandry.

Secondly, I know the common edible or house onion and thirdly, the sea onion or leek.

In very early times the onion was used as a seasoning for food as well as eaten raw as nourishment.

Not only, however, is the onion a seasoning and a nutritive article but also a purifying and curative agent and of great value to man.

Used as a seasoning in food it improves the digestion, has a good effect on the kidneys and action of the bowels.

It is, I think, because of these properties of the onion that it is so much used in hot countries and among all Jews.

I regard the onion above all as a means of improving the digestion, but I warn people against eating too much of it which often proves harmful. One may object that there are people who eat raw onions as nourishment. To this I reply that such people generally eat the onion in combination with hard black bread or with other rather indigestible food, in such case the onion assists digestion.

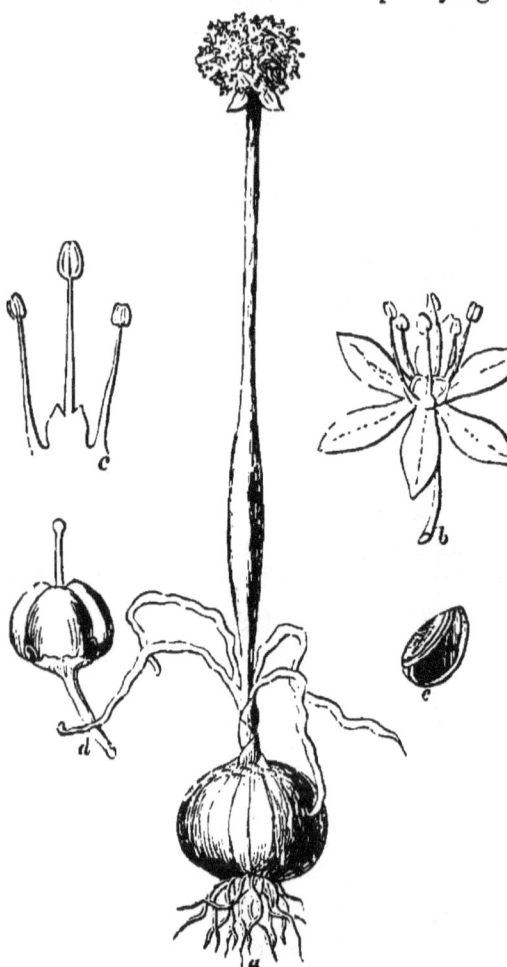

The onion. (Allium Cepa L.)
a) The whole plant, b) The blossom, c) Male pollen vessels, d) Fruit bulb with stalk, e) Seed.

Regarding onion as a seasoning it depends largely upon how much one is accustomed to. There are people who like pepper while others have the greatest dislike to it. Many eat fennel with pleasure while others are sick if they only smell it: Anyone therefore who is a great lover of onions can use it stronger, both as a seasoning and a curative remedy than one who dislikes it.

I would not recommend the eating of raw onions as a vegetable. They are sometimes roasted and eaten after meals instead of a small glass of brandy or liqueur and in my opinion as an addition to the food.

As brandy assists the digestion and disperses the gas by creating a greater stimulus for the system, so do the roasted onions, only with the last there is the disadvantage of a strong odour attaching itself to the eater and which even perfumes the perspiration.

Onions have an unusually sharp smell so that those who peel them get their eyes filled with water and can scarcely get rid of the smell from their hands.

The onion is useful, wholesome and healthy in the household but it may be used as a curative remedy in various ways. One obtains an excellent medicine by cutting an onion into small pieces, bruising it and putting it in spirit. In case of a fit of indigestion or flatulence or constipation, ten or twelve drops of this tincture taken three times a day will be excellent in its results.

Nor are these drops less efficacious in quenching thirst or for people who nurse the sick and are exposed to infection.

Chopped up onions boiled in milk is an excellent remedy for colic and stomach ache. In this form the onion was used many years ago by housewives for children suffering from worms as the creatures cannot endure the smell. Three spoonsful of the liquid morning and evening is a sufficient dose.

Onions can also be cooked with honey: this form of taking onions is extremely good for people who have difficulty in passing urine; three or four spoonsful of this honey and onion-water two or three times a day will be found sufficient to remove the difficulty.

Onion and rosemary boiled together in half water and half wine is also good for the same trouble and an effective remedy in dropsy.

Onions boiled in milk or honey make a capital eye-wash.

People suffering from swollen or running eyes should wash them with this decoction three or four times a day; by doing this the redness and soreness will gradually disappear, the water will be drawn out and the eye itself braced.

Onion juice, or the skin of the onion laid on, is an excellent remedy for a bad bruise or for lumps behind the ear or on the body. In the one case the pain is lessened and in the other the lumps disappear.

Rubbing with onion decoction is as a rule good for every sort of eruption if water applications are used at the same time — without the latter the eruptions would disappear for the time only and on the least provocation break out again.

Onion tea is good for all foul, putrefying sores.

The onion may be used also for other purposes, for example, if you want to colour Easter eggs yellow, cook onion skins in water and dip the eggs into it; they will come out a beautiful yellow colour.

Onion decoction is also the best means of cleaning gold or silver lace.

All these properties of the onion were known many hundred years ago and it is due to our modern way of living that its use as a curative remedy has been forgotten or set aside as unsuitable.

Think how easily the onion can be cultivated! It increases in the garden like a weed and will grow in every kind of soil though it prospers best in a good one.

Sixth Part.

Advice in Case of Small Accidents.

Instantaneous Help for Accidents and Practical Directions.

Preface.

Accidents frequently occur in daily life either in the workshop, kitchen, street, or other places which though, perhaps, at first of slight importance may be fraught with serious results.

This being so, I do not consider it unsuitable to give my dear readers a few directions what to do in various emergencies.

Fractures.

What can a novice do until the Doctor comes if an accident results in a fracture of the arm or leg, before the patient is carried home or to the hospital?

In answer to this I say he can put on an extempore bandage.

But how can one know if a fracture exists?

The unlucky person must be examined and attention paid to the signs of an existing fracture.

As a rule this is recognisable externally through the clothes by the altered shape of the limb, also by the severe pain at the smallest movement or change of position of the affected member together with the hanging down of the broken part all of which are indications of a fracture or broken bone.

A fracture of the upper part of the thigh is harder to recognise than a fracture of any other member of the body. Very closely fitting garments as for instance, boots, stockings, tight trousers must be cut away at once before swelling sets in, otherwise their removal will be difficult.

A broken leg without a wound.

A broken leg with a wound.

The many people crowding round must be made to stand aside and only those who are necessary in helping may remain.

Place the patient in as easy position as possible and then look round for materials with which to make an extempore bandage or splint.

♥ Extempore splint for a broken leg formed of bits of wood and the bandage of pockethandkerchiefs.

Extempore splint for the calf of the leg made out of a flower pot trellis.

In searching for materials one must take into account the scene of the accident whether it be in inhabited places or in the open fields. It may be that you will find in the surroundings rulers, pasteboard, laths, little planks, or flower pot trellis from which to make splints and support for the broken limb.

It is possible to use all sorts of things in the making of extempore splints and for fastening them on, one is glad of sashes, neck cloths, pockethandkerchiefs, sheets, garters or string.

Extempore bandage of travelling rugs or straw mats.

If the accident occurs in the open country one must take the garments of the victim himself for bandages. The entire material must first be prepared so that the broken limb is not raised oftener than necessary.

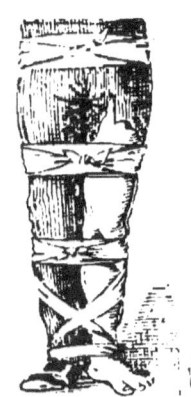

Extempore splint in the region of the knee with a stick and an umbrella.

Bandaging by binding the broken leg to the sound one.

Extempore splint of cooking spoons.

If suitable material is not to be found, then bind

Extempore splints of pieces of wood or any existing instruments and bandages on the fore-arm.

the broken leg to the sound one, or the injured arm to the breast so firmly that a movement of the injured limb is impossible.

~~~~~~

The Transport of Accident Cases.

When the bandage is put on, some method of conveyance must be thought of.

Extempore bandaging of splints, bandages and cloths on the arm.

One either tries to form a litter or get a carriage on which to place the sufferer and so convey him carefully to where he may find surgical aid.

If, however, he cannot be moved without injury, leave him if circumstances permit on the scene of the accident until the arrival of the Doctor who will himself decide on the next step.

Since we are on the subject of transport I will use the opportunity and speak more in detail upon it.

For transport it is usual to employ litters or transportable invalid baskets. Where these exist by all means use them; the question is, suppose you find yourself

Coat or cloak used as a litter.

Straw rope litter.

in the country or in a place where such things are lacking, what is to be done?

"Necessity" it is said "is the mother of invention" and certainly a practical man will know what do to in the emergency. He would make extempore litters of ladders, chairs, window shutters, planks, doors, benches and anything at hand.

In order that the sufferer may lie easily pads must be placed on the hard articles and for these straw, hay, pillows, mattrasses and beds will be best.

Or hammocks may be made of quilts and sheets: as the illustration on this page shows you take the sheet or the quilt and tie it firmly to two poles or you may make the hammock in the same way with cloaks or with two coats; in the latter case you thrust two poles through the sleeves turned inside out of two coats and button them together which at once forms a litter.

You can also make litters of straw bands or ropes, hurdles and straps. If the accident happens in the forest one can make litters of stems of trees and stretch quilts, sheets or hurdles over them.

If, however, only one person is present at the time of the accident he should if possible fetch help from the nearest house or village.

Transport by means of hammocks.

If the patient can walk, try to get him as far as possible, or if there is no danger in the movement or the delay he might walk himself to the nearest place in order to obtain a carriage or other help.

A practical method of carrying the patient over very uneven steep and rough roads is the so called mountain sledge, which consists of two long tree stems one end of each is carried by a horse, ox or cow whilst the other ends slide along on the ground. Between the

13*

stems some covering is laid or rather stretched on which the patient is placed.

The mountain sledge.

This sledge may be used also on even roads with advantage; I am told that the work of transport goes on much more quietly than in a carriage.

Transported by means of a chair.

These last unless well padded jolt too much especially on bad roads.

But suppose there are no means of transport at hand, let some of the helpers join forces and with two pairs of hands make a seat for the patient, he steadying himself by putting his arms round the shoulders of the bearers, or if the man is unable to do this the helpers may carry him with one pair of hands, and with the other pair support him round the loins.

It is very necessary that sick people should be carried with the utmost care, for by this they are saved much pain and an increase of their sorrow avoided.

In these days when nursing the sick is a very prominent occupation, special instruction is given in rendering first aid to the injured and how they should be carried.

How can one save oneself and others from drowning?

If a man who cannot swim falls into the water he should try to get on to his back with the head bent backwards and the mouth directed upwards, at the same time draw in a deep breath and send a very short one out, so as to pump the lungs in this way full of air: the arms must be kept under the water, otherwise the person will sink directly. In consequence of the existence

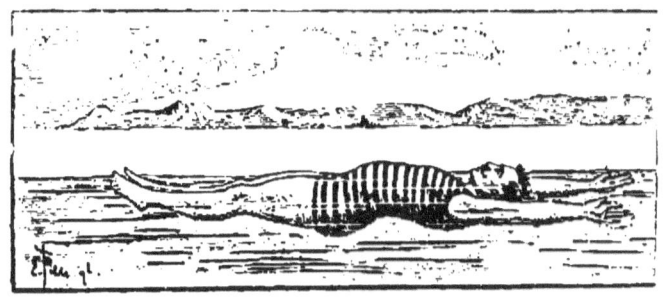

Floating on the back to save oneself from drowning.

of air in our lungs and bowels, swimming is rendered possible, for the air always ascends and balances equally our bodily weight.

If both arms are stretched out behind and above the head the body takes a level or horizontal position, the face and mouth remaining out of the water. If, however, one holds both arms downwards naturally the

I. Wrong position when in danger of drowning.
II. Proper position of the body for the prevention of sinking.
III. Saving a drowning person

IV. Saving a drowning person.
V. and VI. Saving a person who has fallen through the ice

lower part of the body becomes heavier; by lowering the feet the whole body assumes a horizontal position. The head must be bent backward so that the mouth stands out of the water. By slightly moving the hands and feet this is easily effected. If in fright the person stretched his arms up out of the water and called for help, his body would sink and his head go under the water. I would advise every one to learn swimming for by it he may be able to help others as well as himself. Think of the proverb "that which is learnt is not difficult to carry".

It is, I know, extremely difficult to keep one's presence of mind in a case of accident; and in one's anxiety, the most purposeless movements are made which hasten rather than save from death.

It is almost impossible to save a person who has no knowledge of swimming — not only so, but the man trying to help is often himself in danger from the convulsive clinging of the drowning person. It behoves him therefore, to be careful not to be so hampered that swimming is rendered impossible for him.

The rescuer must try to grasp the unlucky person so that the latter's back lies on the former's breast and in that way swim to shore: He should remember never to swim against the stream, otherwise he would be powerless before he could render aid.

How to remove the water from the mouth of a drowning person.

If the victim has already sunk, try to bring him to the surface and endeavour to restore him to life, if he has not been already too long in the water.

If among the crowd of spectators there is no person who knows how to swim, throw ropes to the drowning person, or articles of dress tied together, or oars, poles, life belts, indeed anything at hand which he can grasp and so save himself.

If a person falls through the ice push out to him poles, planks, or ladders, or a long rope with a stick tied on to it horizontally. If you must go yourself to the rescue try to creep to the broken place on your stomach or walk, with a long stick pushed through the arms at your back, slowly to the scene of the accident, if then you fall you cannot sink under. In every case provide poles or sticks.

If then you have rescued a man from such an accident and if he has been a long time in the water or is unconscious so that he gives no sign of life try artificial respiration with him: this must often be continued for hours before it is successful.

Of course it is understood that a Doctor should at once be sent for.

Artificial Respiration.

This is an imitation of the real thing and causes the expansion and contraction of the chest in order to allow fresh air to enter into the lungs.

Artificial respiration.

Lay the patient flat on his back so that the chest is raised, and to ensure this lay something under it;

then place yourself behind him, take both arms by the elbows, raise them gently above the head and hold them firmly there for two seconds. By this the chest of the apparently dead man is expanded and if the air tubes are free, fresh air is provided in the lungs.

Then draw the arms back in the same way and press them on the chest for another two seconds. This movement presses the air out of the lungs.

If two people are helping, each one grasps an arm making the movement simultaneously. This experiment is made fourteen or fifteen times a minute and continued with perseverance until a voluntary and independent respiration is observed.

As a rule the colour of the face alters on the commencement of breathing.

Another experiment is that of pressing the chest flat. Lay the apparently dead man on his back, push a bolster under his loins and cross his arms under him.

One person now kneels above the head and draws the tongue forward from the mouth, he holds it fast by means of a cloth in the right corner of the mouth or pushes forward the lower jaw with his hands laid behind both corners of the jaw.

Another person kneels astride the hips of the unconscious man, puts both hands spread out flat on the lower wall of the chest, draws his elbows to his sides and bends down slowly till his head and that of the victim nearly touch; in this way the person leans forward with his whole weight on the chest of the victim thereby pressing the air out of it.

After this, the person helping, again straightens himself quickly taking his hands away so that the chest may once more expand itself.

This, like the former experiment, must be carried out quietly and regularly.

With children or with weak and emaciated people another method may be employed, viz., that of grasping the body under the ribs with curved fingers from above, thus causing the chest to heave and subside. In these endeavours to restore respiration great care must be taken to avoid haste and violence.

Directly signs of independent respiration set in stop the artificial experiments and try to restore the circulation of the blood and bodily warmth.

Wrap the body of the sufferer in dry garments and rub the limbs thoroughly from below upwards, then put him in a warm bed and cover him with hot blankets, place hot water bottles on the abdomen, under the shoulder blades, and the thighs, and also at the soles of the feet: they must be hot but not too hot and india rubber bottles are best.

When at length life is so far restored that the victim can once more swallow, warm drinks, such as tea, wine or liqueurs are poured down his throat by spoonsful.

Suffocation.

One often reads in ths newspaper of people dying of suffocation.

One or more perhaps have descended into a pit or mine and become unconscious and, because there was no one at hand to bring them out, they never awoke from their faint.

If a man on descending into a pit becomes unconscious it proves that the air is foul and full of danger, therefore, quick to the work of rescue!

Try to procure at once ladders and ropes, and do not forget a protective bandage for the mouth, such as a cloth dipped in vinegar and water.

Before descending to fetch the victim try to get rid of the gas in the pit; this can be done by agitating the air, viz., by shooting, by letting down an open umbrella and drawing it up again rapidly, by shaking water into the pit or by throwing down burning straw or paper, in doing which, however, great caution must be used so that the rescuer is not singed for lighted gases blaze quickly upwards.

The person who now intends to descend into the pit to fetch up the unconscious man must firmly secure a rope over his chest and shoulders and a signal line on one hand so that he can signal if he himself is in danger of becoming unconscious, or else to draw up the victim when he has secured it round him, and, as I have said before, let him tie over his mouth a large cloth dipped in vinegar and water.

Rescuing a man who has fallen into a pit.

Those standing above, of whom there should be at least three or four, must hold the ropes. That to which the rescuer is fastened must be tensely held: the one who holds the signal line must exercise every power of perception he possesses in order to note whether the corresponding arm, on which the line is tied, still moves voluntarily, if on a call from time to time the answers are not distinct.

At the first sign of loss of consciousness in the rescuer he must at once be drawn up.

Should he, however, reach the bottom safely he must seize the unconscious man as fast as possible, tie the second rope to him and give the signal for both to be drawn up.

When the unconscious man has arrived at the top, take off all his tight fitting clothes, wash his face and chest with water and vinegar, and brush the soles of his feet with a rough brush.

Until the arrival of a doctor perform artificial respiration as was shown in the chapter on drowning.

Not only are we in danger of suffocation from foul air in mines, pits, and shafts of wells but are in danger of like misfortune from breathing in carbonic vapour and carbonic acid.

It is not at all of rare occurrence that suffocation is caused by charcoal stoves, escape of gas and foul pipes.

In places where large masses of people meet and where there is also a lack of currents of fresh air, foul air is generated, causing people to faint and where a fainting fit is long protracted it may end in death.

In cellars where new wine or beer is fermenting one runs the danger of suffocation, or at least of unconsciousness by inhaling foul gas, and if help does not quickly come the person affected may die in convulsions. In such a case one must act quickly to be of use and the first thing to be done is to get the person into the fresh air; and here again caution must be used that the rescuers do not themselves become victims.

If one must go into places filled with foul air try at once to produce a draught by opening the doors and windows: or if necessary, the windows may be broken from outside, but let no one enter without first tying a wet cloth over the mouth, then let him take a deep breath and jump in and open every window in the place.

Should an escape of gas be going on, lighting up must be forbidden; rather grope about in the dark. Get the victim into the fresh air at once, and proceed to deal with him as directed for similar cases.

If you are about to rescue a hanged person, use caution when cutting the rope so that the body does not fall down; but hold it and let it slide as far as possible.

Do not place a drowned or hanged person head downwards, or the blood would rush with too great force to the brain and cause probably a shock to it. Lay the person horizontally on the ground, if he has been rescued from hanging, but if from drowning, lay him on his side, the mouth turned downwards, or lay him over your knees on his stomach and hold his head so that the water can run out; see illustration, page 199.

If the mouth is covered with mud cleanse both it and the nose and try to keep the mouth open so that fresh air can penetrate.

Should a person be in danger of suffocation by swallowing too large a piece of food, keep his nostrils closed, put your index finger deep in his throat, and try to bring out or press down the obstacle. If this does not succeed, press the chest and stomach of the sufferer against something firm, and give him with your fist violent blows on his back and shoulder blades so that by shaking about the fixed-in morsel it may be rendered movable, and either ejected or swallowed.

In every case send for a doctor, and explain the accident so that he may bring the necessary instruments with him.

Frost Bite.

What is called frost bite does not only occur in intense cold, but in moderate cold if at the same time by walking too long, or by hunger, or by too much use of alcohol, the person becomes exhausted or stupified, and sits down in the open air and goes to sleep or loses consciousness.

Snowed-in people are more easily brought back to life because snow is a bad conductor of heat. Frozen people are pallid and cold over their whole body except on their hands and feet as well as nose and mouth where a bluish tinge appears; pulse and breath seem to have stopped.

The limbs are stiff and numb, and at their extremities frozen hard and icy cold.

With frost bitten people the work of restoration and warming must be proceeded with quite gradually and with the greatest caution.

First of all carry the sufferer into an enclosed place which, however, must be absolutely cold and unheated, and undress him cautiously which is best done by cutting the clothes from the body so as to avoid the breaking of the stiff limbs.

If snow is obtainable, cover and rub the body violently with it; if not help can be given by cold wet cloths, or cold sand, or a cold bath.

At the same time try artificial respiration. When breathing has set in and the limbs are suppler, bring the patient into a temperate atmosphere, cover him with cold coverings and the room may be made warmer by degrees. In addition to this rub the frost bitten man with cloths made warm gradually.

When animation has returned give strong smelling remedies, such as ammonia or onions, and internally wine or soup.

This same treatment may be used in frost bite of any single part of the body. If, however, this part remains numb and blue and swells up there is great danger of its mortifying.

Burns.

One makes a distinction between burns, scalds, and corrosives. Burns arise from the action of intense heat whether of flame or of molten or liquid metal on the skin and the organs lying beneath it.

Scalds arise from hot water or steam, and corrosives from chemical materials, such as acids, or ley. These three kinds of burns have the same result.

There are three degrees in burns; first, the inflammation of the upper surface with painful reddening of the skin : second, the formation of blisters and third, carbonisation in which black scabs form.

The causes of burns are many and various, namely, explosions, outbreaks of fire, incautious handling of dangerous and inflammable articles and suchlike.

It cannot therefore be superfluous to remind every one to use caution in such things. If the clothes of a person are burning or smouldering, the first thing to be done is to extinguish the flames. Throw him on the ground and roll him over and over, so that the pressure of the body extinguishes the flames.

Unfortunately presence of mind is often lacking in the burning person and he runs wildly about thereby causing a draught, and increasing the flames. Seeing this wrap him round with a quilt, cloth or cloak, or indeed anything large and thick you can put your hand on, and roll him round on the ground or cover him with earth or sand, after this, pour water over him from head to foot.

In scalds one proceeds in the same way by throwing cold water over the scalded person, and wetting his body and clothes thoroughly through. This being done the sufferer is laid on the carpet or table in a warm room until medical help comes.

The clothes above all must be removed with the utmost care and tenderness; they should be cut with scissors or a knife so that they fall off the body of themselves. Never tear or pull them off lest the blisters should be torn.

Blisters and wounds of burns.

One may prick the blisters with a clean needle if they are very much distended so as to allow the water to run out.

If the patient is extremely thirsty give him some warm drink such as tea, for in burns the bodily heat decreases.

To mitigate violent pain use a remedy which will decrease it and preserve the burnt parts from the action of air and dirt, such for example, as ointment, powder, oil, peeled raw potatoes, flour, cream and linseed oil mixed with lime water. It is advisable to cover these remedies with wet cloths and fasten them with a bandage.

In severe burns or scalds, patients are often very quiet, feel but little pain, are very thirsty and sigh a great deal which last is regarded as a bad sign.

In burns caused by acids you may, as in burns, throw cold water over the victim, then put on cold wet compresses.

More particulars on burns and frost bites may be found in the seventh part of this book.

~ ~~~~~ ~ ~

Fainting fits.

These are caused by fright, pain, exhaustion, severe loss of blood, hunger, thirst, tight lacing, poison or disease of the brain.

Since it is often difficult to distinguish the kind and origin of fainting, medical aid should be summoned.

Until this is available, the following directions may be of service. Remove all clothing that confines the body such as collars, neck-ties, belts, dress buttons, stays and such like in order that the restricted circulation of blood may again be freely promoted and give the patient fresh air.

Lay him on his back with his head low if his face is pale but, on the other hand, high if the face be red.

Should the patient vomit hold his head on one side so that his breathing is not affected.

If one can detect no breathing in the fainting person institute artificial respiration.

Not infrequently epilepsy or falling sickness is the cause of loss of consciousness. Where this is the case the limbs twitch convulsively, the face is distorted, the mouth foams and the teeth often bite the tongue.

One cannot do more than prevent such a person from injuring himself; put a soft cushion under his head and wait quietly till the fit is over. On no account try to control the convulsive movements, or unclose the clenched hands or you will increase the convulsions.

~~~~~~~~~

Sunstroke.

This is a dangerous kind of loss of consciousness brought on as a rule by great heat combined with physical exertion such as a long military march under a burning sun.

The sufferer at first feels very tired and giddy; he experiences oppression of the chest, heat in the skin, rapid and weak pulse, laboured breath, dryness of the tongue and violent thirst; the voice is husky and slow and the hearing very weak.

The full effect of the sunstroke may be warded off, if, when these symptoms appear, the man is allowed to rest, loosen his clothes and drink some water.

But there are times when sunstroke causes a man suddenly to fall senseless to the ground; his face gets deep red, his eyes become fixed, his breathing very rapid and rattling, the pulse violent and the skin burning hot.

In a case like this help must be given at once or stiffness of the whole body, foaming at the mouth and even death may result.

The help should consist in carrying the man to a cool place, for instance, under the shade of a tree or a house, and lay him on the ground so that the upper part of the body lies high; then open the several articles of dress, obtain fresh air for him, sprinkle him with cold water, lay cold wet cloths on his head and chest and give him some water to drink.

If the breathing is very disturbed institute artificial respiration and do not spare the rubbing of his limbs.

Finally give him a little stimulant such as wine.

Poisoning.

Poisons are materials which exercise a disturbing effect on the body and even on life. Some, like mineral poisons, affect both stomach and abdomen and cause vomiting while vegetable poisons cause stupefaction.

In poisoning cases give at once a suitable antidote; for acids, give alkali, such as, for example, soda, chalk, potash and the like; on the other hand for alkali poisoning the antidote should be acids such as vinegar and lemon.

For sharp pungent poisons take, for the protection of the throat and stomach, slimy or oily remedies like oil, milk, water and flour, eggs or sugar-water. Further try to get the poison out of the stomach by means of vomiting which can be effected by drinking luke-warm water or milk or some other emetic.

If stupefying poisons have been swallowed every care must be taken to keep the man awake; give him a stimulant such as strong coffee or tea and place cold bandages on the head and mustard plasters on the stomach and employ cold douches.

All these things should be done while waiting for the doctor.

Lead Poisoning.

Glaziers, potters, painters tin and metal workers, porcelain manufacturers and indeed all hand workers who have to do with lead run the risk of lead poisoning.

Lead poisoning shows itself principally in abdominal pains and constipation which often assume such large proportions that the sufferer is rendered incapable of work. The body of the man so suffering is greatly reduced and it is necessary to strengthen his whole system as well as get rid of the poison, therefore he should take emetics such as warm water or milk or warm bark of oak decoction and at the same time take two half baths in the week and drink wormwood tea.

Nicotine Poisoning.

Anyone who has brought nicotine poisoning on himself by his immoderate passion for smoking, should, as soon as he is made aware of his condition, give up smoking entirely from that day forward. The nicotine poison should be ejected from the system by applications of water.

14*

Blood Poisoning.

If the blood poisoning be but slight use a hay flower poultice (decoction of hot hay flowers) which should be very hot and renewed every quarter of an hour.

For severe blood poisoning, take soaked hay flowers, press them lightly out, lay them on a warm coarse linen cloth and place the cloth round the affected part of the body.

The hay flowers should be as hot at the patient can bear them, renew this cloth every twenty minutes and continue the treatment for about six hours; should this first trial be unsuccessful repeat it yet oftener.

Poisoned Wounds.

These arise from snake bites, the bites of mad dogs, poisoned weapons, knives and nails, or by poisonous dyes: The poison may reach the heart from the wound and thus poison the whole blood.

In a case like this bind the affected limb above the wound quite tight with a girdle, band or cloth; then try to get the poison out of the wound either by sucking, if the lips be clean, by cauterizing or by means of an acid such as carbolic acid or saltpetre acid should one of these be at hand.

For a wound caused by the bite of a snake drop sal ammoniac into it and give the sufferer alcoholic drink and get the doctor to the patient as soon as possible.

For inflamed parts of the body omit the strapping up.

For insect stings caused by gnats, bees, wasps and flies dab the affected part with some drops of spirit of sal ammoniac and try to remove the sting. Tincture of arnica is equally good to drop into the wound.

For severe inflammation apply thick loam compresses: they should be rather wet so that they do not get too quickly dry. For all sudden accidents use such remedies as you know will afford quick help and prevent death.

When immediate danger is over one can institute regular treatment and authorized water applications.

The application of various bandages.

Bleeding.

In all cases of wounds veins are injured which accounts for bleeding.

The sort and amount of bleeding depends entirely on the size of the vein which has been hurt. If small veins only be injured the blood simply flows slowly and weakly out of the wound. Such bleeding can be at once quieted by pressing together both edges of the wound which causes the running blood to draw together into a sticky clot: it is possible sometimes to check the bleeding by holding the injured limb in a perpendicular position.

If larger blood vessels are injured the blood flows out dark red and far stronger than in an injury to a small

Injured pulse artery.

Broken varicose veins.

vein, especially if pressure on the limb above the wound is exercised.

Here one checks the bleeding by the removal of all tight-laced articles above the wound and by the raising of the limb and light pressure thereon.

If, however, an artery be hurt the bright blood spurts out in jerks and in a strong jet causing danger to life from bleeding.

Until medical aid comes bring strong pressure to bear on the wound or on the main artery above the wound.

The wounded limb is raised up so that the severe outflow of blood is

Pressing together the jugular vein or carotid artery.

Pressing together the main branch of the arm-pulse-artery.

lessened and the wound and limb bared as far as the trunk.

After this clean folded linen is laid on the wound and pressed by the hand or bandage tightly against it. Should the blood flow on in spite of this, press the wound tightly together above the pulse artery.

One can press lightly together the pulse arteries which lie close to the surface, as, for instance, in the upper arm, thigh or neck.

This treatment of pressure with the fingers requires however anatomical knowledge, practice, skill and endurance.

Therefore it is better to bind the injured limb up with an elastic bandage and this not merely once but

Pressing together the arteries of the thigh.

Pressing together a pulse artery in the upper arm.

Pressing together of the large arm pulse vein.

Pressing together the arteries of the arm by means of bits of wood and bandages forming a Tourniquet.

frequently so that at last the pressure of the binding is so strong that the blood can no longer flow through the veins and so the bleeding stops.

This is best managed by application of Tourniquets and elastic braces, supposing these should not be at hand a linen bandage so arranged that one edge overlaps the other will suffice.

Fasten the ends securely and sprinkle the whole bandage with water so that it will contract tightly of itself.

Braces as a temporary means for pressure on the veins.

Another method is that of folding an ordinary large pockethandkerchief loosely about the limb like a cravat round the throat and pushing a stick under it and twisting it round until, in consequence of the resulting pressure, the blood stands still (See illustration page 216) yet this sort of pressure should not last too long otherwise the whole of the bound limb might become gangrenous.

In all these methods of checking the flow of blood great care must be used.

Before resorting to tourniquet try to check the blood by holding the limb up and pressing on the wound and vein; should this remedy fail proceed to the tourniquet.

These remedies are however only meant for first aid until a skilful doctor is on the spot.

Exercise great caution also with the application of so called popular remedies such as cobwebs; if these for instance, should be taken from a dirty corner of the house and laid on the wound the dust would easily penetrate and so increase instead of lessening the mischief.

Neither should novices use chemist's remedies such as chloride of iron etc.

Tourniquet on the thigh by means of pieces of wood and bandages.

Sudden Seizures: Epileptic Attacks.

It is not an unusual thing for people to be attacked by sudden illness in the street and for those passing and repassing to stand and look on because they know not how to help.

The right thing to do would be to get the sufferer quickly into a house or some quiet spot and give him a glass of water and allow no crowd around him. The same course should be pursued with epileptics.

I strongly advise those who know that they are never safe from attacks to go out as little as possible alone! It is not safe: and want of care in this direction has brought many an added misfortune on the sufferers.

Injuries to the Eyes.

Should an accident happen to the eyes by the squirting of acid into them the best way of helping would be to wash them rapidly with fresh spring water.

It may happen that in the loading or unloading of chalk or lime some may get into the man's eyes. What is the best remedy to apply in this case?

My answer is **sugar** which is to be found in every house.

Dissolve two-thirds of an ounce of sugar in one ounce of pure warm water; get it cold as quickly as possible and with this decoction wash the eyes . thoroughly until the burning sensation passes from them.

The same may be accomplished by strewing powdered sugar on the eyes, still I prefer the dissolved sugar if there is time to prepare it. After either of these applications the eyes should be washed with cold water.

Bits of iron of various sorts and sizes which may have found their way into the eyes may be removed by a magnet and washing them thoroughly with cold water.

Wounds from Bites.

The best remedy for bites of dogs and injuries of a less serious kind is careful bathing of the wound with cold water and rubbing in diluted tincture of arnica.

For bites of mad dogs compare "**poisoned wounds**" on page 212 with "delirium" in the 7th Part.

Falls.

If a person has the misfortune to fall into a chalk pit the simplest thing to do is to wash away the chalk by a heavy douche and so prevent further danger.

After a fall on ice or slippery roads, streets or steps, cold water compresses and rubbing with tincture of arnica, diluted with from three to four parts water, give the most relief.

Wounds occasioned by the fall require above all to be cleansed by washing with fresh water; having done this staunch the blood with arnica tincture and bind up with a compress.

If a fracture of the knee pan (patella) or any other bone has taken place make a compress of cold water at once which will prevent the injured part from swelling so much, that the doctor on his arrival cannot feel or prove it.

Rupture or Hernia.

The disposition to rupture is as a rule inborn and excessive mechanical force is but rarely the cause.

Ten per cent of men are afflicted with ruptures which have partly arisen in childhood and partly in riper years frem various reasons.

A rupture may be caused by the incautious bearing of heavy weights or by coughing violently when standing; by great exertion in preaching and by screaming especially among children.

My advice is that every one whether his constitution be good or bad should, when carrying or lifting heavy weights, keep his mouth shut and his legs close together.

My advice to preachers and to all who speak or recite much is to wear a belt.

When coughing violently in bronchitis, influenza and inflammation of the lungs the person should if possible sit down or lie on the side and draw the feet up or press the finger on the vicinity of the groin while coughing especially if inclination to rupture exists or is actually there.

Anyone having the misfortune to get ruptured should go to a skilful surgeon at once and have proper bandages or trusses.

Many put these on the bare body, but I do not recommend this because of the foul smell which collects from the excretion of the body.

Rather wear the truss or bandage over the shirt where it can be kept quite clean.

If the rupture occurs in childhood there is a chance of outgrowing it if suitable bandages, douches and baths are made use of, but with older people this can scarcely be expected. Still the evil should not be neglected, on the contrary every means within reach should be made use of for its alleviation.

Now and then it is good to lay on the injured part a cloth dipped in bark of oak decoction; and cold water will also greatly relieve the sufferer.

All other remedies, though much vaunted, may be considered useless.

Struck by Lightning.

When the electric current strikes a man it has not always a fatal result: therefore relieve him at once of all burdensome garments and wash him rapidly with cold water. Artificial respiration should be at once resorted to if he is unconscious.

Colic.

What is to be done if one is attacked by violent colic in the middle of a journey?

People liable to such trouble should never stir from the house without having a remedy in their pockets. The best is brandy, a teaspoonful for every attack.

When one reaches home, if the pain has not completely gone, go to bed and put a poultice of warm camomile decoction on the abdomen.

Dislocations.

A dislocation may easily happen either from a blow, or push or by any forcible movement.

By dislocation we mean the moving of a bone out of its proper position a circumstance which causes great suffering to the individual.

a) The head of the Humerus in its proper position.
b) The head of the Humerus pushed out of place by dislocation.
c) Elbow in the proper position.
d) Elbow pushed out of position by dislocation.

While waiting the arrival of the doctor lay cold compresses on the dislocated part.

Sprains.

If a person sprains his foot what is to be done?

The best thing is to go to bed and bind on a loam bandage, that is a bandage dipped in cold loam water, and renew it every half hour until the sprain goes.

The bandage should never be allowed to get hot; as soon as it is warm it must be renewed.

If the foot is bandaged in this way immediately after the sprain occurs, further remedies will scarcely be necessary; if, on the contrary, the foot is neglected serious consequences may arise.

Seventh Part.

Diseases.

Mens sana in corpore sano!

Diseases.

Egyptian-Eye-Disease.

The Egyptian-Eye-Disease, which has its habitat principally in hot lands, is both infectious and painful. The eyes are inflamed and the longer the inflammation lasts the weaker grows the sight.

Without entering more minutely into the causes and symptoms of this illness I will briefly describe the quickest way to cure it.

It is very necessary here, as in most other diseases, to operate upon the whole system if we would remove the special mischief.

If the eyes are inflamed poultices of pot-cheese are of great service. By dilution a delicate salve is made out of the pot-cheese and laid on the naked eye without putting the salve on a strip of linen.

The poultices which should be renewed three or four times a day are very cooling to the eyes and remove the inflammation.

A drop of clean, pure honey dropped into the eye every third day is an equally good remedy for cleansing it.

We have, however, yet another remedy to offer which is an eye bath of wormwood decoction given alternately with one of fennel water.

Still the pot cheese is, according to my judgement, the most efficacious in this disease.

For benefitting the whole body I recommend full douches, half baths, thigh douches and back douches, one or two daily according to the constitution of the sufferer.

As soon as the corrupt matter has been got rid of from the body, the eyes will get better; but it is a process which demands patience as it is only gradually that the system can be made healthy.

Inflammation of the Eyes.   Opthalmia Syphilitica.

A new born child was attacked on the second day by a bad eye disease. Both eyes were frightfully inflamed and swollen; and the inflammation increased daily until at last the child was blind.

The two doctors called in for consultation could think of no way of helping and gave up the case as impossible of cure because the disease was of a syphilitic character.

Then the child was brought to me and I felt no doubt that with proper treatment it would get quite well again.

The whole day I had poultices of pot-cheese-water made and placed on the eyes fresh every hour; and once in the day I had the child dipped in cold water. Very soon the inflammation abated and the swelling yielded and in the course of a few days the child was cured.

The child is now five years old and has always enjoyed good sight.

Small Pox.

As long as I can remember the law has existed that everyone must be vaccinated.

The first vaccination takes place in childhood, the second towards the end of school days, and lastly men are vaccinated on their entrance into military service.

For years the controversy has raged between the adherents and the antagonists of vaccination. I am completely convinced that vaccination is quite unnecessary, and I ask, wherein lies the purpose of it?

According to the doctors vaccination is intended to chase all impure matter out of the human body. For that, my dear readers, vaccination is not in the least necessary seeing that hydropathy renders this service far more effectually.

Anyone who understands how to use water properly will not find it difficult to remove everything that is unwholesome from his body.

Of the bad consequences resulting from vaccination I could give many examples, but of what use would that be, seeing that the law exists and must be obeyed if we would escape punishment.

Now let us speak about Small Pox which may be divided into black and white, dangerous an less dangerous.

Small Pox plays very little part in hydropathy as water acts in all cases only alleviatingly and bracingly.

I recommend instead of vaccination and application of cold water on the whole body.

If Small Pox is in its first stage it is good to operate as much as possible with water.

I should like to cite a few examples of patients suffering from small pox.

A few years ago a division of soldiers was quartered in the Dominican Nunnery near here. In spite of

15*

all of them being vaccinated, as they confided to me, they were attacked by small pox and in such a malignant form that even in a few hours five men had fallen victims to this insidious disease.

On the whole about twenty people in the building were laid low by small pox: Two nuns were so rapidly carried off by death that one could not even be sure of what illness they had died.

Nor was it confined to the nunnery; it spread so rapidly through the village that whole families were lying low with small pox.

I myself exhibited every symptom of it and the doctor assured me that in three days at most I should be stricken down by this disease.

In order to avert the danger I took from that moment a half bath every hour.

When I had taken several baths I felt better, the fever and thirst yielded and at the end of eighteen hours I felt a great improvement.

My lost powers were recovered and after another six hours I was able to leave my bed.

Anyone who knows the effect of water and understands how to apply it properly will not fail to obtain help.

The symptoms and fore-runners of small pox are lassitude and weakness, violent headache and fever, loss of appetite and sleeplessness; these, together with excessive languor, are constant attendants of this disease.

During the time I lay ill in bed many others in the village were attacked by small pox and my first action naturally answered for these patients.

Where the illness was in its first stage I proceeded exactly the same as in my own case and indeed with equal success.

If, on the contrary, the small pox had made further progress I ordered the patient to put on daily a shirt dipped in warm hayflower decoction for the purpose of obtaining an increased extraction of diseased matter and to counteract the strong increase of the small pox.

In addition to this precaution a half bath or a whole washing must be given three or four times.

I also tried bandages in some cases but I found that my first method of the shirt dipped in the hay-flower-decoction was the simplest, surest and most effective.

If this treatment is continued for ten or twelve days it will remove the disease without fear of relapse.

As regards internal treatment my principle is that where no appetite exists a person should not be forced to eat; and drink, in any case, should be given in small quantities. Beer, spirits and wine I never recommend; the most I allow is a little wine mixed with water because the patient easily turns against water alone.

Preserved fruit diluted with water forms an especially good cooling drink: Honey-wine or what we call **Mead** I also recommend.

Very often I order some kind of tea such as worm-wood and centaury or tormentilla of which the proper dose is a spoonful every hour.

Even after the illness is over the treatment might be continued; certainly the whole washings daily and the half bath should not be omitted as they will help to make the health of the person stronger than before.

Cholera.

This is such a dreaded disease that the mere pro-bability of its presence terrifies people and no wonder.

If we cast our eyes back on the past and note the sin and struggles of mankind we are bound to exclaim "the dear God knew best when He let loose great and fearful epidemics among the sons of men."

When cholera has once found entrance in a place it spreads with such frightful rapidity that in a short time whole families, households, villages and even towns are laid low by it.

I believe there is no more infectious disease than cholera. It may be conveyed by the air, by articles of clothing, by mutual intercourse, and indeed by many and diverse ways and methods.

Whenever and wherever it appears it creates general fear and horror among the people, for it rages everywhere and demands the lives of many. In my office as Chaplain I have also had opportunity of learning something of this frightful disease.

In the evening a person going to bed quite well and healthy would be awoke after two or three hours by the pains of cholera and dead next day: so fearfully rapid is the development of cholera.

The symptoms of the disease are abdominal troubles and noises in the body followed by violent diarrhœa; and on reaching its climax severe and constant vomiting is added. The sufferer is devoured by thirst, urine ceases to pass, he experiences convulsive pains and an ever increasing restlessness sets in so that he tosses about in bed almost without a restful moment.

The course of the disease varies; in some cases it proceeds quickly, with others slowly. If the skin of the patient becomes cold and the cheeks fall in, if the pulse gradually gets weaker and a clammy sweat breaks out on the forehead, if the voice fails and diarrhœa increases then death is near.

At the time cholera raged in Swabia I was able to save many from its fatal ending.

The best remedy is to keep the patient very warm so that the whole body properly perspires and not to give him food that may cause disturbance and produce diarrhœa or increase it.

Guard also against taking a chill; and I advise not only the patients but also the healthy to wear an abdominal belt.

What I have prescribed up to this holds good not only for patients but also for the healthy; for in times when such a fearful epidemic is abroad the greatest prudence is necessary.

The most suitable food during the reign of cholera I consider to be water-gruel and barley and to avoid all stimulating and spiced nourishment: and the best beverage I believe to be tea or red wine mixed with water.

We will now pass on to the special treatment of a cholera patient.

First I take a cloth folded several times and dipped in hot water and lay it on the patient as hot as he can bear it, in order to get the system as warm as possible: Internally I give him peppermint tea.

In about half an hour I renew the hot damp cloth

This in a short time will produce an intense perspiration and in a few hours the pains will yield and. the crisis will have passed.

After every third bandage I give a whole washing and internally fennel cooked in milk. Two or three spoonsful taken warm every hour greatly comforts the patient.

Colic is treated in just the same way as cholera for both are malignant and require equally to be taken in hand at once.

If the former is neglected it may become extremely malignant. No matter in what form these two diseases set in one cannot do better than put on at once the warm bandage.

Epilepsy.

Epilepsy is, not without reason, one of the most dreaded of diseases. People afflicted with it scarcely dare to mix in the society of their fellow creatures.

In my youth this disease was rare and the person afflicted with it was regarded as incurable; the older the person the stronger and more lasting were the attacks and it was no unusual thing for the person to die in one of the fits, often indeed in the open air where he fell on his face and was suffocated. According to the length of time the disease had power over a man so did his mental powers decrease till a complete derangement took place and left its mark on the face of the sufferer.

Epileptics had a fixed staring look, the complexion was red and blue and they were corpulent and dribbled at the mouth.

They had also an extraordinary appetite eating sufficient for two or three people which accounted for their well nurtured bodies.

For the rest they were rather insensible to joy and grief, worked half unconsciously and with difficulty were kept at it.

Their life in short was a painful one and rendered more so because they were feared and shunned by everyone.

I think people fled from them because they were afraid that witnessing an attack might produce the disease in themselves but of course this fear was unfounded.

There are several kinds or degrees of Epilepsy: First there is the actual, acute Epilepsy in its severe form and in its lighter form which last does not make any great disturbance in the system and is not so noticeable externally: then there is Epilepsy with many of the characteristics of hysteria which is more easily cured. The disease in its acute form is reckoned among the incurable.

The seat of Epilepsy is probably in the brain on the coating of which it is surmised that a diseased irritating matter has formed.

A large number of sufferers from this acute kind of Epilepsy fall suddenly without any warning and become unconscious; a convulsive twitching of all the limbs sets in, the eyes become fixed, foam issues from the mouth, the complexion turns blue, the breath rattles, and anyone witnessing the seizure for the first time thinks that death must certainly be near.

Thus the patients lie for a longer or shorter time; then by degrees they become quiet and recover consciousness, the blue tint vanishes and gradually they attain to their former condition except that for some time they feel weak and suffer from headache.

These attacks at first come on gradually perhaps three or four times in a year; in time, however, they increase in strength and appear much more frequently so that at last several attacks may occur on the same day.

The system succumbs to these too frequent attacks, a proof that great disturbances must exist in the organism and that the whole body is undermined.

In the acute form of Epilepsy there are sometimes signs that the attack is close at hand; for example the patient becomes giddy, tries, without success, to say something, endeavours to support himself but staggers round in a circle and then falls to the ground.

Formerly it was a common practice to force open the mouth of anyone overtaken by an attack with a key, people thought that the rattling was a sign of choking and that aid could be rendered in this way.

Just so, they also tried to force open the fists of these poor people because they clenched their fingers convulsively: they rarely succeded nor was the effort at all necessary.

For if the patients could not breathe through their mouths they could do so through their noses and beside this Epilepsy does not affect the breathing organs at all.

The best thing to do is to get the sufferer into as comfortable a position as possible and let him rest. The attack will cease of itself and in this way he will be best helped.

If the person attacked is wearing tight clothing and if the neck is wrapped about, the coverings must all be loosened so that air can be admitted and the burden on the body lessened.

Epilepsy of the less severe sort cannot be reckoned among the decidedly incurable diseases.

The invalids generally have warnings when an attack is at hand; they feel giddy and have as they say a feeling as if the whole system was upset and in haste they seek some corner where they can sink down; all is dark before them and they lose consciousness.

In this condition they remain for a longer or shorter time until gradually they regain consciousness.

They remember nothing of what has just happened but they feel headache, exhaustion and general weakness which in time disappear and the whole system returns again to its former condition.

Others afflicted with this species of Epilepsy have no warning but may hinder the attack by a sudden movement or by rest. It may also occasionally be warded off by speaking to the patient quickly or by taking his hand and pressing it.

Again there are others who stand rigid as if overtaken by Catalepsy completely unconscious as though petrified: they do not even move an eye; they remain in this state for a short time and by degrees come to themselves and continue their occupation.

Yet again, there are those who in falling to the ground draw up their fingers, arms, and legs convulsively similar to sufferers in gout. This condition however does not last long and when it is over the invalid feels as he did before.

Others again are attacked during the night in bed: they know and feel nothing about it and only learn it from those who are in the same room with them or who observe it from seeing that their bed is more disordered than usual.

And finally others have attacks when illness or special eruption is lurking in the system and will not come forward: they know that an attack is coming but cannot make any resistance and therefore succumb.

If the root of the mischief in these people is cured then they become perfectly healthy again.

The third kind of Epilepsy has many symptoms which bear a certain resemblance to hysterical convulsions. If these hysterical predispositions are removed the attacks cease. Such people have attacks on the smallest provocation: A slight alarm or anxiety, an outburst of anger, any trouble, and above all anything which suddenly affects the temper will bring them on, therefore all provocation, where possible, should be avoided.

As regards the treatment of the first kind of Epilepsy in which the mental powers are already weakened and the circulation of the blood has become greatly disturbed, the only thing to be done is to try and render the position of the sufferer easier: and even when the attacks become weaker and less frequent one can do no more than this. In this and in every kind of Epilepsy it is necessary to use only gentle, non stimulating applications.

If patients suffering from the acute form of Epilepsy rise from bed to take two or three whole washings weekly and a half bath twice in the week these will be found sufficient.

In all acute cases there are larger or smaller disturbances in the blood and system generally therefore we should set to work to get these into a healthy condition.

I advise the invalid to wear once a week or once a fortnight a shirt which has been dipped in hay-flowerwater and to take every week two whole washings and a half bath. A short bandage in addition would be of advantage.

Many Epileptics have an eruption on the face, or sores on the body and they will derive much from taking a head vapour bath once a fortnight in addition to two whole washings and a half bath.

When all the impurity has been drawn from the system the condition of the invalid will be greatly improved.

For internal treatment we must give only such food as can be easily digested, that will not create much gas and that will promote the formation of blood and ensure a general bracing of the system.

If the invalid is very weak give him daily a little bone powder as much as will fill a salt spoon; this will have a good effect. For the improvement of the blood, give him wormwood and sage tea or rosemary and wormwood.

Such acute cases of Epilepsy as I have described above used to occur in my youth but nothing like so frequently as now.

The increase both of Epilepsy and insanity is immense.

Formerly the Asylum Irsee for the insane was quite sufficient for the Swabian and Neuburg district. Now a much larger one has been built in Kaufbeuren and yet is quite insufficient.

So also Epilepsy has spread in such a fashion that the unfortunate sufferers are found in large numbers in every period of age: and if one enquires for the reason of this frightful increase it is traceable mainly to the modern mode of life viz, the mistaken nourishment, the great debilitation and the unreasonable clothing.

I am firmly convinced that the increase of Epilepsy is connected with the decrease of the average duration of life of man.

Certainly the germ of the disease may have lain in the grandmother: but in the mother it develops further and in the child it becomes a regular disease.

There can be no doubt that the unsuitable nutriment, the mistaken clothing and the wrong mode of life have done much to develop this disease, whereas a reasonable mode of life and regular bracing would have done much to stamp it out.

This supposition is confirmed bythe fact that this disease is much less frequent and severe in remote places lying far from towns, and where the old simplicity in nourishment, clothing and mode of life still obtains. Wherever we find this disease rampant it may be traced to the circumstances just related. If we desire energetically to combat with it we must first prevail on ourselves to lead a more reasonable life and revert to a good simple nourishment and clothing as well as bracing.

In this way the disease will be best attacked and beaten and even if it should still appear in individual cases it will always find the body capable of resisting it.

Epilepsy of the second or less severe kind embraces even a larger number of cases than the first, but it is capable of cure. Still one must not forget when curing it that the diseased stuffs have carried on their mischief for years in the body and cannot be got rid of in a moment.

As soon as we know the immediate causes of the attacks we must naturally try to remove them as quickly as possible; for example if an attack comes on after excessive indulgence in food or drink we must guard against all excess; if it is produced by mental excite-

ment, avoid it as much as possible; if on the other hand
excessive heat or cold acts unfavourably we must protect
ourselves from the one or the other.

To obtain a cure we must brace our systems and
take good, simple, nourishing food.

It is better to take small portions often than a great
deal at long intervals which latter causes overloading of
the stomach.

Much fluid such as wine, beer and similar beverages
are certainly harmful.

The clothing should be simple and capable of protecting the body from the inconveniences of the weather
and not occasion disturbances in the circulation of the
blood.

The oftener and the severer the attacks so much
the simpler and gentler must be the treatment. Thus
for instance walking bare foot on wet stones once or
twice a day for a quarter of an hour conduces greatly
to bracing and at the same time draws the blood from
the head. In addition to this take a whole washing
from bed.

If these simple remedies are continued for several
days the invalid may, instead of walking on wet stones,
walk in water for three or four minutes as high as the
middle of the calf. Beyond this he can take a thigh
douche daily and wash the upper part of the body.

When the patient has done this for several days
then I advise him to take one day a thigh douche, the
other a half bath beside washing the upper part of the
body daily. The half bath is especially effective for these
people.

As further treatment one can take a half bath one
day and a back douche the next and walk once daily
in cold water or on wet stones. Both full douches and
back douches are very effective.

By these applications the body is brought out of
its feeble state and the whole system becomes stronger
and more capable of resistance.

By way of specially nourishing and simple food I advise for breakfast malt coffee cooked in milk, and for a change strong broth a little thickened. Oat-soup and others of like character are equally good but they must not be too highly seasoned or salted, otherwise they act as irritants on the nerves.

I am in favour of all that is good if it be simple, pure and unadulterated, therefore the invalid may choose for his midday meal any food that can be easily digested and that contains good nutritive matter.

In the evening I recommend good strong broth or some milk food, or a little meat.

I must, however, observe that food composed of refined flour has very little nutritive value and therefore pure flour should be used. Anything cooked in milk is easily digested and nourishing.

Various remedies may be taken internally with good results. The best will be those which promote a good digestion, operate strengtheningly on the body and at the same time extract the diseased matter.

Epileptics frequently have much gas in the body which operates unfavourably on them, therefore remedies should be at hand for this.

I have found lavender-oil especially effective against the gas, taken daily morning and evening, the dose being from five to eight drops on sugar or in water.

It is not good to confine a patient solely to one medicament for a long time, it is better to make a change; so in turn with lavender-oil use fennel or oil of cloves. Tea of bark of oak and wormwood and sage also operate favourably on the stomach.

Very good results have been obtained by taking in the morning a spoonful of oil of cloves or some other delicate oil which strengthens and warms the stomach and therefore is a protection against attacks. Another protective remedy against the attack may be found in taking care of the abdomen.

Hysterical convulsions or attacks pass readily into Epilepsy according to appearances although the cause of each is quite separate.

As regards the care and cure of these sufferers the same treatment holds good as that given above for the second species of Epilepsy.

Proper applications' of water, good food and bracing are as valuable here as elsewhere.

Obesity.

In nearly every place one comes across certain people of whom it might be thought that they belonged to some totally different race and sex so fat and shapeless are they.

They generally possess a good temper and good appearance and nothing about them gives the impression that they lack anything.

We must not imagine, however, that these people are as extraordinarily healthy as their fresh and prosperous appearance would lead us to suppose.

Not a few of them complain constantly of many infirmities and get angry because they are only laughed at. Their walk is slow and burdensome and with their stiff feet they can only take quite short steps; their breath readily yields and if they attempt to talk while walking the machinery stands still, their strength fails them and in short they appear to one like a butcher who carries about fifty or sixty pounds of meat and cannot dispose of it.

Their eyes are prominent, their face is puffed out and spotted blue and yellow, even their lips and ears are sometimes blue: their stomachs are like over filled sacks, their arms are very fat but the flesh is flabby and loose.

These conditions are visible also in the thighs which are fat and stiff so that it is impossible for such people to pick up anything from the ground.

Generally they are most phlegmatic, the result of their physical infirmities; all enterprise is lacking in them and they prefer to vegetate rather than to live a full vigorous life.

Everybody is subject to obesity, but women seem more disposed to it than men.

Corpulent people as a rule take a great deal of fluid such as beer, tea and wine and they not only drink but eat largely; but we must confess that there are country people who live on the simplest diet and are moderate in the amount they take who yet are fat.

This condition of the body may be inherited just like other evils; I have known people who tried to avoid it by eating and drinking as little as possible, but it was of no use: in spite of their self denial they became as round as a ball.

It is worth while to ask here "What is the chief cause of obesity?"

The answer is I think that corpulent people have poor blood which, instead of strengthening the muscles, creates fat: this reacts upon the blood which suffers both in quantity and quality and prevents the development of the natural powers of the body.

In such a condition fatty degeneration of the heart and internal organs sets in which not only hinders their action but increases flaccidity so that by degrees the whole machinery stands still and life is brought to a close.

Corpulent people are also subject to obstructions in the chest and lungs because the organs work so laboriously that the proper excretion of used up matter is not possible, thus one mischief helps the other and the condition grows gradually worse.

If the action of the heart decreases, the feet usually
swell and as a consequence of disturbances in the circulation of the blood the kidneys become affected and
other obstructions form which may end in dropsy.

Because the whole machine is overladen and the
organs weak from excess of fat, corpulent people suffer
occasionally from apoplectic seizures and sink under
them.

On the other hand the digestive organs are generally in good order and one rarely hears complaints of
them from a very stout person.

The same symptoms and results may be seen among
animals.

The question arises, may one take means to reduce
or prevent the ever-increasing fat without fear of injury
to the system, and if so, how should one set about it?

It is a mistake in my opinion to take certain prescribed medecines for it, for with these the patients are
forbidden to take their usual diet, but on the contrary
are ordered to take one with which they are not at all
familiar. The system accustoms itself with difficulty to
a new and strange diet; the vegetarian attacks violently
a meat diet, and vice versa, and the result is not happy.

In addition to a strange diet purgative remedies are
given to corpulent people, consisting mostly of aperient
mineral waters.

Under these circumstances the system is in a difficulty; it has to digest an entirely strange diet and, having done this, remedies are given which carry off the
nourishment digested before it can be of benefit. It has
therefore a very hard and difficult time of it.

This treatment will neither strengthen the system
nor reduce the evil satisfactorily.

True the corpulence may decrease but the weakness
will increase: the good digestion ceases and the system

becomes spoilt and ruined rather than strengthened and braced.

A man twenty-nine years of age sought me out and made the following statement, "I have undergone a six weeks' course of treatment for corpulence and have only lost one pound: Can I be made thinner here?" On my answering in the affirmative, he at once said "Then here I remain".

On giving him directions he asked "what diet must I keep to?" I replied, that he was not to eat little but well.

My reply annoyed him and he said "Sir, I do not allow myself to be jested with: you must not imagine that I am a glutton or a drunkard; I have suffered a great deal these last six weeks from hunger and thirst and if I were not in earnest I should not have come here".

I answered "I was not joking; use the applications and a very nourishing diet and you will be again all right".

With that the man was content, followed the cure for six weeks and then sought me out again.

"It is inconceivable to me" he said "that I did not lose more during the six weeks in which I suffered hunger and thirst, and now that I have eaten and drunk as I pleased, yet not in excess, for the same length of time, I am lighter by twenty-nine pounds and feel fresh and strong instead of staggering about like an old man."

How did this happen? .

During the period in which he suffered hunger and purgative remedies the corpulency did not decrease although the system became inactive.

If, however, it had decreased the body would have become weak at the same time.

16*

By my plan viz, the water cure, the whole body
was strengthened, all flaccidity and superfluous fat were
thrown off; the applications effected a stronger and more
active interchange of matter and thus the whole system
was renewed and freshness and strength of life and purpose were reinstated.

**What problem has the hydropath to solve when he
wishes to reduce the size of a corpulent person and
what is his remedy?**

The corpulent person is always a weakling and
requires strengthening; therefore such applications must
be used as will endue the whole organism with strength.

The body of the corpulent person is spongy and
flabby and contains much superfluous matter; therefore
the hydropath must operate on the ejection and rejection
of the same so that the organs may return to their
normal condition. On the other hand the nourishment
necessary to the body must not be removed because this
is absolutely needful for the regeneration of the whole
system; and because it is only when the organs work
properly that a rapid interchange of stuff and the removal and ejection of the superfluities are rendered possible.

Therefore a diet must be recommended which gives
good blood and supplies the body with power.

The corpulent person should avoid fluids as much
as possible except where actual thirst exists and he
should not take food which causes thirst.

Look at the animal world; an ox, for example, goes
in harness and works just as long as he receives dry
food: If, however, he gormandises on grains and generally wet food his strength diminishes, he gets difficult
breathing, becomes fat and is only fit for the butcher.

The flesh also of a stall-fed cow is spongy and
flabby and never so strengthening and sustaining as the
flesh of one which goes in harness and works.

The applications prescribed by the hydropath to the corpulent man for the strengthening of the body are thigh, back and full douches; while the half bath, the short bandage and the Spanish mantle operate specially on the extraction and purging of the superfluous matter.

An official who weighed three hundred and twenty pounds came here (Wörishofen) for the cure. His head was full and swollen, the neck almost as thick as his head, his abdomen unusually large and arms and thighs equally enormous.

He enjoyed drinking beer and wine and indulged besides in a very generous diet.

We gave him daily two applications; one to draw together the spongy, powerless organs and strengthen them; and the second to brace and reanimate the whole body.

The applications consisted of three half baths weekly, two thigh douches and later on two back douches beside two or three full baths. The food was chosen according to the rules given above.

He was accustomed to beer and wine therefore they could not be suddenly denied him without causing harm; so he was allowed a small quantity of both one and the other until he at length renounced them. The improvement was very marked; each day he became fresher and more vigorous; his sleep improved, his appetite increased and his mental condition kept pace with his physical till, as he himself said, he began to live over again.

If the lightning douche can be properly applied by one who understands it the corpulent person would derive great benefit, but the greatest caution is necessary in giving it. For if the patient were very nervous, a strong lightning douche would only increase his nervousness.

Therefore instead of the lightning douche we more often recommend a full douche together with a short bandage, or a lower and upper compress, or a Spanish mantle once or twice a week.

All these applications dissolve and extract.

I reject all internal remedies in corpulency unless they conduce to a good digestion and promote the formation of blood.

Two cups of tea each week made of tormentilla, angelica-root and wormwood to be exchanged later for tea of bark of oak, sage and ribwort will be of great service.

Fistula.

If often happens that from time to time a sore appears on the body: At first it may be only a speck somewhat inflamed but as time goes on it spreads, the inflammation increases, and it becomes a large sore from which more or less matter flows.

These sores occur not only on the surface of the skin but also in the interior of the body either in the vicinity of a bone, or in the muscular region, or any other place.

Naturally the diseased matter seeks for an outlet somewhere: it forms a canal through which it flows and is carried off, although externally we see no more than the opening and the discharge.

Such an internal sore with an external outflow of pus is called a fistula.

The discharge may last for weeks, months, and even years and affects the adjacent parts of the body so that bone may be eaten away or muscles greatly injured.

Fistula is therefore in many respects like an internal disease and may, under some circumstances, cause death.

For, like other internal diseases, it weakens more and more the organs attacked by it, and gradually draws the whole body into sympathy; and because of loss of power and lack of good wholesome material it is impossible to maintain strength in the body which daily becomes weaker.

If the outflow from the fistula ceases before the internal sore is healed the mischief within rapidly spreads and the invalid feels much worse.

As long as the fistula runs, the patient feels tolerably well, but as soon as the outflow stops and the fistula closes, symptoms appear of many and burdensome diseases.

The cure of fistula is very difficult because its cause must first be removed and we all know how extremely difficult, if not impossible, internal treatment is.

My belief is that water is the only means by which the entire body can be operated on so as to throw off all diseased matter, strengthen the system, and heal the sores from the inside.

The general effect of the application of water is certainly not to hinder the outflow nor on the other hand to promote it.

When all the diseased matter has been removed from within the fistula heals of itself.

The question then is how is one to set about the cure?

A man came to me who had for years suffered from a **rectum fistula**; his case was considered hopeless because the fistula was thought to be malignant.

I prescribed for him applications of water on the whole body specially however on the abdomen and for this I ordered thigh and back douches, half baths and sitting baths.

The effect of these was the gradual strengthening of the whole system, a diminishing of the discharge as well as of heat and pain: then we applied clystier or what is known as "injection" to the wound, the first day with a decoction of pewter-grass, the second day with one of bark of oak and the third day with one of fenu-greek.

These remedies cleansed the wound and increased the man's bodily strength till at length he became quite well.

Such cures of rectum fistulas have been numerous here.

In the same way we cured other fistulas, if they had not gone too far to derive benefit from injection or **irrigation.**

The chief treatment must always be on the whole body to enable it to get strong enough to throw off all foul and putrid matter, for when this is accomplished healing begins.

Herpes.

Just as in houses dust and dirt accumulate in every room and corner so does diseased matter find its way into the human system: this matter, which has been caused by obstructions, gives the system a great deal of trouble, for it can neither make use of it nor eject it.

Thus it frequently happens that severe eruptions appear on various parts of the body and spread over the whole surface and are known by the name of Herpes.

These eruptions are not all of the same character and are generally divided into three classes; viz., into **the running or wet herpes** (Eczema), into the **dry** or **scaly herpes** (Psoriasis) and into the **spreading herpes** (Lupus).

Wet Herpes. Eczema.

Eczema is an eruption which, beginning in one place, spreads gradually and may even attack the whole body.

The blisters, which are at first small, break in time and a sharp fluid runs from them. This outflow is uninterrupted and a general spreading of the trouble sets in. If the air gets to this exudation it becomes dry, and thick crusts or scales form.

From earliest times small children suffered from these exudations and housewives tried all imaginable means to cure the bad heads.

Children afflicted in this way were otherwise quite healthy, probably on that account, because the diseased matter was thrown off and ejected. Certainly when a child escaped this form of herpes it was feared that it would not turn out healthy.

I look on this outbreak as a **natural vaccination** where the system is itself strong enough to eject the unhealthy matter.

As in children this disease appears on the head and face, so adults suffer with it on various parts of the body, indeed the whole body is sometimes attacked: it may be produced by the trade or profession or by the locality in which one lives or works.

The shoemaker, for example, gets eczema in the hands and the weaver on various parts of his body if the place in which he works is underground or lacking in good sanitary condition.

A distinction is made between Eczema suffered by children and that suffered by adults and the treatment for each should also be different.

Among the remedies employed by mothers for healing these eruptions on the heads of their children was a salve made of marigolds and many prided themselves in making a perfect cure by means of it. Others made a salve of sorrel which was highly thought of. Thus many household remedies were used not only by the mothers but also by the doctors.

The children often suffered for months from this eruption and carried traces of it for years on their heads and faces and even when the scars seemed to disappear they were not always quite healed but would break out in some other form.

According to my conviction, and I have had years of experience, **this treatment was not right.**

The pores were closed by the ointment and the diseased matter remained in the system where it developed into some other form of illness: even it washings were applied to the diseased parts they merely caused a crust to form on the skin and the diseased matter could no longer get an outlet.

It appears to me exactly like a person who, anxious to catch mice, stops up the holes; for a short time there will be peace until the mice have found other outlets or made fresh holes.

If the eruption repeats itself again and again it will come out in the eyes causing weak sight, inflammation of the eyes and even cataract.

This clearly proves that even in the childish system there exists a great deal of diseased matter which must be ejected from the whole body if the child is to be healthy.

This can only be done by strengthening the system so that it ejects the impure matter and forbids other to enter, and the only means by which this can be effected is by the cold water treatment and a few other remedies.

Warm water would not answer at all, it would only weaken the system and develop the diseased matter more rapidly.

The children, instead of being bathed in warm water, should be dipped daily in cold water but not for longer than a couple of seconds.

If they be weak and sickly dip them only every other day: yet I must say that I have had the weakest and most sickly children dipped daily with the greatest success and never saw harm from it: the plunge should be as short as possible and the head need not share in the dipping.

To effect the ejection and absorption of the diseased matter from the child's system put him on a shirt dipped in warm hay-flower-water which will both extract and absorb the diseased matter.

It is necessary to remove the diseased matter as rapidly as possible from the skin, otherwise it eats corrodingly round about. By means of the shirt the matter is drawn from and out of the head.

If the supply of bad matter to the head ceases a new skin forms under the crust and the mischief is cured.

If the head is not bandaged it should be washed once or twice daily in fresh cold water as the sensitiveness of the child permits.

If it is desirable to hasten the cure yet more a decoction of pewter grass may be used instead of fresh water for washing the head.

As this eruption is very biting and burning children try to get relief by **scratching** which naturally increases the evil.

As soon as they are dipped in water or have the hay-flower-water-shirts on, all scratching ceases.

Under ordinary circumstances a shirt twice in the week will suffice, but where a child is strong and the eruption bad three shirts in the week may be given for

the first fortnight.   After this twice, and finally once in
the week.

Even when the child is considered to be cured of
the eruption and restored to health the cold water treatment should not be discontinued.   For even when all
appearance of herpes has vanished, time is needed to
complete the internal cure, and at least for three months
the child should be dipped in cold water if not daily, at
least two or three times a week.

In this way the child will become steadily stronger
and healthier and the diseased matter internally will
also disappear.

I see no use in giving tiny children tea but when
they are bigger an internal remedy may not be out of
place; and for cleansing the blood and improving the
digestion ribwort tea may be given.

Yet I value above tea variety and change in the diet.

It is a principle which may never be disregarded
that the greater the variety of nourishment the easier
it is for the organs to take what is necessary for them:
and further that uniformity in nourishment has no good
results either with children or adults.   If for example,
the child has been living on diluted milk for some time
give it some black malt coffee mixed with honey.   This
serves to cleanse and at the same time provides a good
deal of nourishing matter.

As we do not recommend pure milk for children a
change is all the more advisable.

Wet herpes is harder to cure in adults than in
children, because all the impurity which has collected in
the system presses out day and night in individual spots
which eat up the skin and form themselves into a crust
which then dries, falls off, and is replaced by fresh.

The evaporation is often almost unbearable, and even
if the eruption is not particularly offensive yet the matter

absorbed in the clothing causes a sickening odour so that the necessity for great cleanliness and frequent change of clothing is most important.

As regards particular treatment I know of no remedy beside cold water. Of course it is possible to take something which will improve the blood, get the digestion into a better condition, and give the whole body more strength. The used up matter would otherwise remain too long in the body and dissolve itself into impurity and eat away the skin. It is plain therefore that one must operate on the whole body in order that the foul matter may be ejected, the stomach improved and proper digestion established.

I advise as water applications whole washings and half baths together with thigh and upper douches.

Commpresses may be applied but not too many as they extract often so rapidly as to increase the weakness. People with strong constitutions may use the Spanish Mantle once or twice a week but scarcely oftener.

Experience has convinced me that douches in combination with wet shirts and Spanish mantles answer best. It is especially beneficial if the shirt or Spanish mantle be dipped in loam water and kept on the body for a time not exceeding an hour.

If the patient is strong this application can be made twice or even three times in the week, otherwise once a week will be enough.

I consider change of food especially important for operating internally.

In selecting a tea let it be one that will both cleanse and strengthen, such as bark of oak and sage, the first heals and cleanses and the latter acts upon the chyme (food reduced to an even, fluid mass).

Wormwood, Centaury and ribwort all act beneficially on the stomach and digestion.

As regards the discharging sores I have found it
best to bind a small cloth or piece of linen round them
which has been dipped in pewter grass water and renew
it frequently: the effect of which is to remove and absorb the excretion and hasten the cure. I have also
frequently dipped the cloth in loam water taking care
to renew it constantly.

The loam*) should be reduced to a fine salve with
pewter grass and tormentilla waters.

A specially effective remedy is to dry the loam or
Fuller's Earth on the hearth or stove, powder it and
dredge it over the running sores.

I must once again remark that the water applications should not be made too frequently in cases of
people suffering with this disease because their systems
are weak. One application daily will in most cases prove
sufficient.

The half baths should only last from one to two
seconds and the compresses should only remain on three
quarters of an hour.

Dry Herpes or Scaly Herpes.

This form of disease receives its name because of
the scales or scabs which form on the skin and fall off
as new ones appear.

The characteristics of this kind of herpes differ from
those of the former. In this, it is not at all unusual
for the whole body to be covered, filling the patient's
clothes with scales and even where he sits or stands the
scales fall about.

This herpes appears generally in large or small spots
for instance one thigh may be quite covered by them
while the other remains free. They show themselves
specially where pressure is exercised on the body as at
the elbows, or at the knees, and so on.

---

*) The word loam means Fuller's Earth.

A sufferer from this kind of herpes may look and feel quite well and think that the disease is all on the outside of the skin, but when he is attacked by gastric or other trouble because the herpes have not come out sufficiently he knows that the disease has its roots not on the upper surface of the skin but deep down in the system.

These herpes are not infectious because the dry scales are not in a position to attach themselves to any other person's skin but they may be inherited and pass from the parents to the children or, skipping a generation, appear in the next.

This disease would not be painful if the irritation arising from it did not produce inconvenience and sleeplessness, but it is a disagreeable burden both to the sufferers and to others because of the strong and often repulsive evaporation from it.

It is to this class of herpes that the scurfy head belongs which is nothing more than the dry shedding of the skin of the head and may occur often not only with children but with adults also.

Even if this kind of sore head is not dangerous it shows that such people have a thin scalp and that the pores are closed by these scales or scurf.

Dry herpes require a long time to cure because to effect this a thorough alteration must take place in the system: and the only means of bringing this about is to operate upon it strengtheningly and extractingly.

If this disease has its origin in the blood, this must be acted upon: many remedies may be made use of to effect an improvement in it partly nutritive and partly curative.

Variety of food is scarcely more necessary in any disease than in this where the blood requires a great deal of usable material. I have found **dry food** of benefit such as good bread quite dry and reduced to powder and so taken. As an internal remedy nothing is better than wormwood, angelica and sage tea: or bark of oak, ribwort and centaury tea.

If the invalid retains his strength while suffering he may take two full douches, two half baths and two back douches in the week and one or two lightning douches if he is able to procure them; very strong people may take a spanish mantle twice in the week but weak people one or two whole washings.

The washings are very effective if the water is mixed with a little vinegar which last braces the skin and cures small evils.

Scaly or scurfy heads are easiest cured by a general application of water on the whole system; the hair itself must be well combed and the head well washed at least twice in the week with pewter-grass decoction.

I have learnt by experience that long and severe mental work often produces scaly or scurfy heads.

Casting our eyes over the chapters on herpes we come to the conclusion that a diseased evaporation is the chief cause and that this shows itself sometimes by dryness and sometimes by discharge. In both cases however the first cause is formed by internal disturbances from the time of imbibing the food until its ejection.

Further we see that by a special and general treatment the health and strength of the body must be operated on in order to have success.

The Beard Herpes.

There are some people apparently strong and healthy without trouble of any kind who yet suffer from various ailments and infirmities but which luckily are hidden by the hair or dress. For example, **beard herpes** is a serious evil and causes the greater trouble because it cannot always be disguised and in some instances quite disfigures the face.

I look upon beard-herpes as an eruption proceeding from the body even as other kinds of herpes. Many sufferers with this particular form of the disease have come to me and fortunately most of them have been cured. One of these told me that he had had the hairs of his beard torn out because he had been assured that only in this way could he be freed from the disease. Yet, as he complained to me, he had endured all this pain without in any way lessening the disease.

He had applied many corroding fluids and salves, indeed everything that had been recommended to him, without experiencing the smallest alleviation of his sufferings.

As in other kinds of herpes I operate by working on the whole body and have up to now achieved the best results.

If the whole body is cleansed and strengthened the beard herpes also diminishes and comes to an end.

I am perfectly convinced that by exercising only a local remedy we cannot get at the root of the disease and the employment of salve in my opinion does more harm than good as it simply stops the discharge. One must work with the purpose of finding an outlet somewhere or other for the corrupt matter.

The local operation should be merely cleansing dissolving and extracting. I have found strong pewter-grass-water the best remedy for beard herpes. Wash the affected parts well every two hours during the day with this decoction and at night lay on a compress of the same and renew it every hour.

Pewter-grass mixed with tormentilla and used in the same way often produces a yet quicker and more effective result.

The applications on the whole body should be as follows; once in the week a head vapour bath which operates dissolvingly and causes a heavy perspiration.

The duration of the application to be about twenty minutes.

After three or four such applications the head vapour bath will no longer be of use.

As further treatment I prescribe during the first week the upper douche, half bath, and full bath about once in the week alternately; after some weeks two half baths and one full bath will suffice.

Internally tea of ribwort or pewter-grass will best strengthen the system. A cup full daily should be taken.

It must not be supposed that beard-herpes exists merely on the upper or outer part of the skin just because the matter has found an outlet there probably started by a razor or pungent soap. With patience one may be freed from the disease in the space of two or three weeks unless indeed it is a case of long standing then it will require a longer time.

If however beard-herpes is healed by the water-cure it is so thoroughly cured that it never appears again.

Devouring Herpes or Lupus.

Lupus means "Wolf" a creature that cannot easily be tamed and which is in general a wild ravenous beast. It is therefore very appropriate to call the corroding Herpes, **Lupus** or **Wolf**. If this herpes is taken in hand resolutely when it first appears it is not difficult to cure: if however it has laid firm hold on the person or if it is deeply rooted in the system then it is very difficult indeed to deal with, for the blood has then become so corrupt and poisoned that there can scarcely be any question of cure.

Is medicine of no use in this disease? Indeed I know not, I only know that it is declared by the medical Profession to be incurable when in this advanced stage.

I have, myself, made many experiments as regards this disease just because it has so often come before me stamped as incurable and I have been successful in my treatment of it.

Lupus appears in many parts of the body but principally in the face, near to the nose, or ear or eyes.

I think that the reason why Lupus appears more often in the face than elsewhere is because it is more exposed to the air than other parts of the body and therefore attracts the blood which, being very corrupt in these people, eats a way out.

The commencement is so small and so nearly invisible that people carry herpes about with them for years without noticing it. It shows itself first in a scarcely visible swelling as small as a pin's head. By degrees similar pimples collect round the first and resemble an eruption. After a time they break and the sharp discharge eats into the flesh round about until half or even the whole face is covered by the disease and hundreds of scars are embedded in the skin.

The mischief continues uninterruptedly and the patient suffers immensely. Sometimes individual spots seem to have escaped and the patient builds hope of cure on this circumstance. Very soon, however, the poison finds an exit on these very spots which begin to swell: if it attacks the nose it becomes twice its proper size, if it touches the lips they become shapeless sores and so with other places in which the disease appears.

The face of the sufferer is often so disfigured that he is obliged of necessity to be banished from the society of men.

In proportion as Lupus is horrible to look upon so is the body rendered infirm by the development of the disease; the colour of those parts of the face not yet attacked by Lupus is yellow red and white, the bodily strength decreases, general weakness sets in, the appetite is bad, the features are distorted and the whole body becomes emaciated.

There are various degrees in this illness which depend upon the diseased matter, that is to say, whether it collects in large or small, hard or soft sores and how

17*

ejected. If a man, suffering from this disease, is left to his fate he gradually succumbs.

As regards the cure it can only proceed on double lines; viz: by uninterrupted treatment on the whole system because the disease proceeds from the whole body and by its effect harms the whole, and secondly by working on the wounds themselves.

As, according to my idea, all discharge from the diseased parts proceeds from the whole body, the whole body must be enabled to eject the corrupt matter; the system must be strengthened so that the disease does not spread, that the circulation of the blood is regulated and that a general strong transpiration may take place.

If the whole body is operated on extractingly the system gradually strengthened and braced and supplied with good nourishing food so that the blood is improved then there is a fair prospect of cure.

For at first and even at times when the disease has made tolerable progress, the powers are still rather good and the patient does not suffer beyond the endurance of the sickly eruption.

In such a case the Water-cure may be applied with great success. During the week give the patient two half baths, two washings of the upper part of the body, two full douches and, if the system permit, two whole washings and one hay-flower-shirt or a spanish mantle.

If the invalid is tolerably strong all these may be given but should his system be somewhat enfeebled reduce the number of applications by one half.

At the same time he must take a good nourishing diet. I have found dry food excellent in such cases; for example black rye bread cut in pieces, dried, ground to coarse flour and two or three spoonsful eaten twice a day. For the rest a very simple bracing food is to be recommended such as will form much and good blood. For the improvement of the blood and creating a good digestion many remedies may be used in combination with good diet and indeed a variety is best.

Thus bark of Oak tea taken in small portions of one or two spoonsful three times a day is very effective; it must however only be used for a short time as it might cause constipation. Tormentilla and wormwood operate on the stomach and the blood: ribwort, coltsfoot and pewter-grass cleanse and improve the blood, so do juniper berries, rosemary and sage.

I desire specially to mention here that one must not take these teas in large portions, the smaller the quantity the better the effect. Take one or two spoonsful only two or three times a day: one should set apart special days and hours for taking the various teas. These herbs may also be taken in the form of powder twice a day about half a salt spoonful at the time.

We can treat the injuries themselves by removing the discharging matter as quickly as possible, and so prevent it from doing further mischief, and also give such remedies as will dissolve and extract the impurity without hindering the cure. And here it is well to make use of a variety of remedies.

Thus for cleansing the Lupus from all dirt and corruption a decoction of pewter-grass, tormentilla and wormwood with which to wash the sores three or four times a day I strongly recommend.

Occasionally the wounds may be dusted with finely powdered Fuller's Earth yet they should not be dried after the washing. The loam or fullers-earth powder absorbs, decreases the heat and improves the wounds. This treatment may be given daily for a short time and then every three days.

Instead of the above named decoction use occasionally one of bark of Oak, and, in place of the washing, dip a soft linen rag in the decoction and make a compress of it renewing it every hour.

If the affected part is rather swollen a plaster of Fenu-Greek alternately with other remedies is much to be recommended. Diluted alum-water used two or three times a week has a very good effect.

In my opinion one of the best remedies is a salve formed of hemp-agrimony and lard. It struck me that if burdock root and hemp-agrimony were such excellent curatives they would be good also for **Lupus** as a salve and I was quite right. After this salve had been smeared on there appeared many little tiny ulcers or sores which were outlets for the internal disease. For some time the salve was gently spread on daily but I must warn those who use it not to be frightened if an increased eruption appears, for this is the best thing that could occur. The corrupt matter must be drawn out, if the injured parts are to be healed: the principal effect must always be to draw from within to without and this is just what happens in pursuing the treatment laid down by me.

Even if Lupus is a fearful disease it can be cured at its commencement and, even when it has made progress, something may be done to effect a cure.

But never forget that if Lupus is long neglected, that is, until it comes to the stage of eruption, its cure can only be attained by long patience and with great difficulty, and one should bear in mind into what Lupus develops if nothing is done to counteract it.

Troubles with the Feet.

A young Woman came to me and told me that her otherwise healthy sister had had such an irritation in her feet that in scratching them she had caused the blood to flow very freely; soon, she said, tiny sores formed which, however, were quickly removed by the salve ordered by their doctor. She felt happy and thankful at finding herself so well and easily cured.

Her contentment, however, was of short duration, for scarcely three months had passed when she experienced a heaviness in her head and oppression in her chest, her former good spirits had vanished as if by magic, and she began to cough; in a word she had got into a very bad condition and the Doctor told her that the lungs were diseased.

I had no doubt whatever as to the sole cause of this miserable condition which was none other than the healing up of the feet. The diseased matter was pressed inward and its outlet stopped therefore it did its wicked work within.

I advised this patient to wrap her feet from the toes up to the knee in a cloth or bandage soaked in warm hay-flower-water; I also ordered her to take an upper washing every morning and a short bandage during the week as she was very stout. She took also in the week two or three half and full baths.

After a month of this treatment the outlets again appeared and the appetite returned. For internal application she took daily a cup of ribwort or wormwood tea.

I cannot sufficiently impress upon my readers that whenever diseased matter forms itself into an eruption on the body one must not try to heal up the wounds or sores, but, on the contrary, operate on the body so that the diseased matter is extracted completely. The cure will then follow of itself.

Mental or Brain Diseases.

If we compare man with other created living things it is clear that he is the master piece of the whole creation, and that the Creator has stamped it with His own likeness. If mind and body work in full harmony one with the other Man can do noble things indeed, for it is in the exercise of reason, sense and free will that man is a true likeness of God.

Just as great as man appears to us when his mind and body are in perfect harmony just so unhappy is the whole man if this harmony is disturbed; for it is not the soul alone nor the body alone which constitutes man but the combination of both.

As, however, the body can be hunted, weakened and hindered in its development by innumerable diseases so can the mind be brought into the most miserable condition by various adverse circumstances.

It is upon this diseased condition of the mind known as **insanity** that I desire to say a few words.

During a visit I once paid to a friend in an Insane Asylum he took me to one of his patients whom I had formerly known; I was startled at the change which had come over him. In life one often sees people get very bitter and angry over some matter; they get into violent passion, curse and swear and so completely lose self control that for the time it is not safe to be near them. Involuntarily our heart is filled with a mixture of aversion, fear and anxiety. But all this is but a trifle compared with the fearful outbreak of rage which I witnessed in that maniac. Great and lasting was the impression made on me by that sight and I pondered deeply as to how such a condition developed itself.

I thought on the patients who, in their feverish state, did not know what they meant, said, or did, whose minds were befogged and bound by mysterious chains and it seemed to me that this condition was the first step of insanity. Then I pictured this first step to myself in its further development and I again saw that unhappy Maniac who, acting as if he thought his life in danger, dared and defied everything and every one.

But the clouding of the intellect does not show itself always in violence, for example there are those who brood about as if in dumb perplexity, as if they were victims to hopeless misery and profound despair.

They have neither rest nor sleep, they associate with no one, they enjoy nothing not even their food, it is difficult to get a word out of them and they pay no more regard to their relations than to any one else.

In this deplorable state they linger until their last vestige of bodily power is exhausted and death gives them release.

From what has been said it may be gathered that some destroy and squander their life-energies in phrensy and mania which condition is called insanity, while others remain in dumb bewilderment and gradual decay, a condition known as melancholia.

I knew a girl healthy, fresh and full of life, well educated and preparing successfully for a professional career. She lived with her parents and had no anxiety.

She began to be somewhat sullen and sulky, then she became suspicious of and averse to her parents and relations; finally, she refused them all obedience and looked on every one as enemies and persecutors who wanted to do her harm and from whom she felt obliged to defend herself; at last it was found necessary to place her in an Asylum when her condition grew worse; where formerly she had only temporary attacks of insanity they now were constant, indeed it came to such a pitch that she would no longer wear clothes and ate her own excrement. Finally she ended her life in this madness.

Another person whom I knew lost a beautiful Estate by some misfortune. This loss so preyed on her mind that she could neither eat nor sleep and neither would she work.

At first she bewailed and bemoaned her lot but after a time she became as if dumb and gave no answer to any one.

Every endeavour to bring her comfort and mitigate her misfortune was unsuccessful, she refused all help and at length died of starvation not having taken even a drop of water for fifteen days.

We see from these two examples how mental disease may develop in quite diverse ways and how it arises from small, almost invisible, beginnings.

As the large tree was at first quite small and by degrees, under various influences, attained to its present dimensions, so also the commencement of insanity is quite small and unnoticeable and no one may presume to declare whether melancholia or acute mania may result.

It is possible indeed that both conditions may appear alternately in the same individual until one, or the other, obtains the upper hand. The tendency to this disease may have been inherited and only attained its full development as years went on, or it may have been called forth by circumstances such as cruel treatment, important or even unimportant losses, unhappiness, lawsuits, enmity, quarrels or business anxieties.

If, in any of these cases, the tendency is not combated from the very beginning the most terrible results will follow.

A daughter wanted to marry a Farmer's son but the parents objected on the score that his farm was not only too small but that it was encumbered with debts. According, however, to other people's judgment the girl had made a good choice and even the parents had no word to say against the man himself, but they wished their daughter to be more comfortably settled. Neither girl nor parents would give way and the months passed on. The girl began to get melancholy but no notice was taken as the parents thought to bring her into submission to their will. However, the melancholy so increased that nothing could be done with the girl. The parents saw their mistake when too late for the girl was beyond help and died in that condition of Melancholia. Of course they were greatly blamed by their friends who declared that the child's unhappiness and death had been brought about by the parents' avarice.

Of examples such as these one could quote many.

As a boy I had a very talented school friend of whom great hopes were entertained. He was scarcely sixteen when he lost his father and from that time he undertook the management of the Estate himself, helped only by a reliable neighbour.

In a few years he brought the Estate into such a flourishing condition that the most prudent Father of a family might be proud of it.

After a time it was observed that in contrast to his former cheerfulness he had become silent and out of humour.

I met my school friend nearly every year and whenever he heard that I was in the neighbourhood he sought me out. One day I met him after not having seen him for a long time and at the first glance I was startled at his shocking appearance.

All youth and freshness had disappeared, his features were altered, almost beyond recognition, and he had lost all his cheerfulness. I asked him what was the matter, what had happened to make him look so ill? His answer was "I know that all is not right with me. I do not feel well but I have no idea what is the matter. I am no saint but I cannot be accused of any special evil which would have produced my present condition, I assure you I have hours in which I feel utterly crushed".

I saw at once that the whole man was sick through and through and I warned his mother privately that she should take the greatest care of him. I was sure that he had great disturbances in his blood and probably a diseased liver and spleen, and I gave her some advice as to how she should treat him.

When we took leave of each other I said "Till our next meeting" He replied "Certainly not in Time any more, it cannot last long with me."

Six days later he desired to go to church to seek comfort and rest and begged his Mother to accompany him. When she had made herself ready she went in search of him and found him hanging by a rope.

A year previously when I visited him he related all
he had been doing and among other things he mentioned
that a woman in the next village had hanged herself.
The way he told the story showed me that this incident
had made an extraordinary impression on him and he
added "After all Life is man's most precious gift on earth
and he tries to preserve it in every possible way; it is
a fearful condition for a man when he no longer values
or loves his most precious gift, Life!"

This pleased me and I thought "Well you have
right noble thoughts and feelings in you, especially com-
passion for the misfortunes of others".

When I heard of what had happened to him and
thought of what he had said to me a year ago I said
to myself that perhaps even then the germ of this disease had been in him.

We see by these examples how this fearful disease
may begin in a small way and develop itself in mind
and body till it so gets the upper hand that the man
can no longer control himself, in fact the disease resem-
bles a flood which engulfs a man without its being possible to save him.

There can be no more important question to answer than this "How does Mental disease arise?" and yet
I am convinced that no one is in a position to answer
this question completely, because not only are the causes
so many but the disease calls forth such strange attacks
that it is not possible to give an explanation which
shall prove satisfactory and exhaustive on all points.
Who, for example, can explain exactly the origin and
working of cholera and many other diseases?

In mental diseases the causes are even more varied
and obscure than in physical ones because the mind
ranks higher than the body and enquiry, however skilful, is baffled by greater difficulties.

Here and there we are able to trace the cause of
a disease brought into the world by children or their

tendency to it. Of one thing we may be sure that many of the diseases from which our bodies suffer are due to our own fault and this is equally true of menta troubles.

We need not be told that immoderate indulgence in alcohol produces delirium tremens; that sin and vice destroy the powers of mind and body; that the loss of temporary possessions may so work upon a man that he falls into apathy and sadness and eventually into insanity; that a senseless craving and striving after the impossible upsets the brain and leads in time to this frightful disease.

Beside these, with which we are well acquainted, there are many other causes.

Although we may know the immediate causes of the disease it does not follow that we can distinguish the manner of its operation on mind and body. One thing is certain viz.: that one must set to work very carefully in the treatment of mental trouble.

As the disease has developed out of small beginnings so must one begin with the simplest and easiest remedies in order to reinstate the harmony of mind and body.

Neither violence nor strength must have any place in the treatment if you do not wish to increase the evil.

The question is pressed upon us, "Should the mentally diseased be put in asylums or treated privately?"

If the disease makes progress and the patient becomes dangerous to himself and others then without doubt the Asylum treatment is to be recommended above all other.

If, on the other hand, the disease has not yet come to its full development the sufferer may be treated in his own home, provided he has a good kind relative within it and the influence of the home is not disturbing to him.

Unfortunately the peace and harmony of the family circle are often greatly disturbed by the invalid's persistence in bringing forward again and again his diseased ideas; especially if his circle of relatives refuse to admit their importance and try by constant contradiction to bring him to another frame of mind. It is just in this direction that parents and relatives make mistakes; for in opposing the ideas of a mentally diseased person they not only confirm him in them but they make him more obstinate and unmanageable and lose all their own influence over him.

A man brought his sister to me with the following statement. "My Sister is mentally diseased; she is quite imbecile, she neither believes anything we say nor has she any faith in the words of our Clergyman or Doctor." On hearing this the girl tried to put herself right with me and said "My brother is the stupidest man, he understands nothing and pretends to be cleverer than I am, yet I must know the matter best and as to the clergyman he is even more stupid than my brother for he says "Give up your ideas and get a new head! But the Doctor is the most stupid of them all who never says anything but "she is a fool!"

All this the girl said in her brother's presence with much violence, then pointing to her chest she continued "Within here I have long had a devil who knows all about me and I know a lot about him; and it is this that I so clearly see and feel and know that they want to tell me is not true". While she rambled on in this way I thought to myself "If I act like the others she will treat me the same way as she has them". So I looked earnestly at her and said "You are a strange girl and it can't be quite as it should be in your inside."

From that time she never ceased raving about the ignorance of her brother, the Clergyman and the Doctor yet treating me with the greatest respect as one believing something of what she said. "Now at last", she answered, "I have found some one who understands me and lets me have my say".

To this speech I replied "If now I really understand your condition as you admit you ought also to believe that I have remedies to improve it." "Yes" she answered "I believe it and also that you can do and accomplish more than the Devil".

"Good" I rejoined "I will give you something and you will get better."

"I will take everything from you" she said "but nothing from the Clergyman or the Doctor for they understand nothing".

The patient received from me prescriptions and regulations as to conduct which she willingly followed and at the end of two months she was quite well.

I advised her brother not to contradict her in any way and if possible to avoid saying anything to her about her illness; and if she should at any time refer to it to lead the conversation into another channel especially towards something cheerful. I further advised him to keep her well employed in the open air if possible, it should be work that she liked and not such as would exhaust her; and above all every circumstance that would excite her to violence must be avoided.

I have often noticed that people force patients such as this girl to fulfil religious duties. This is very wrong seeing that more frequently than not the attempt even upsets and excites them; so one should be wise in dealing with them and never forget that they are mentally diseased! Try to interest them and sympathise with, rather than contradict, their mistaken ideas. Never make demands on them which they in their condition cannot respond to. The girl whom I have told you about got on quite well with her people as soon as they carried out these rules.

It happens frequently, however, that the family and relations have not the gift of managing the mentally diseased and insane and cannot, if they would, nurse them appropriately.

Under these circumstances it is necessary that the patient should . go to other people who have power to gain his confidence and exercise the necessary patience in dealing with him.

Thus a Pastor came to me and said "I have a mentally diseased parishioner whom it would not be good to place in an asylum for if she were compelled to associate with imbeciles I fear she would always remain one. The difficulty however is that her people cannot keep her at home, they have not the necessary patience with her and she has no confidence in them. Formerly they lived in perfect harmony one with the other, now there are constant dissensions. Could she come for a time into your cloisters and work with the Sisters for she is a good needlewoman? I really believe that by degrees she might get all right".

The attempt was made, the Sisters took her to work in their midst, were very cheerful with her and forced her to do nothing to which she was not inclined and so avoided every occasion for excitement.

She was always complaining "I am a lost creature, who has been faithless to God and myself. I cannot be helped". She suffered from melancholia. She became so disturbed that on the third day of her stay in the Cloisters in an unwatched moment she flung herself out of a window on the first floor, happily without sustaining any damage.

By the loving treatment of the Sisters she became quieter day by day. She began to look much better and at length became quite well.

In cases of mental derangement the body requires nursing as well as the mind. I am of opinion that disturbances must exist in the system of one whose mind is unsettled; for as the disease brings fancies and delirium so an ill conditioned circulation of the blood acts disturbingly on the mind.

If a passionate person or what is known as a spit fire is constantly teased and excited his blood becomes disordered and acts unfavorably on the whole system till gradually the spit fire turns into a fool and the harmony between mind and body is utterly destroyed.

Not only do disturbances in the current of the blood cause mental disease by degrees but impure matter getting into the blood will produce like results.

For example, two children ate some hemlock berries; one of them died and the other became imbecile or mentally weak and to such a degree that the mental powers vanished without very great disturbances having set in. The child remained in this condition for more than ten years from which time the mental powers slowly returned and the child became a right useful, industrious person.

In the treatment of mental cases one must endeavour to remove the cause of the disturbance whatever it be and when this is successfully done one may hope for a cure; if, on the contrary, no attempt is made to get at the root of the mischief cure will be impossible.

There is no better remedy for acting simultaneously on mind and body than water. Even though there be no indication as to the origin of the disease one may use water with good effect. We have all witnessed a battue where not only hares and deer fly before the pursuers but everything that can walk or run; something very like this is brought about by the application of water which sets the mischiefs in our system flying. The best thing to begin with is to provide a very wholesome and nourishing diet by which the blood will be increased and strengthened and a more rapid interchange of stuff within the system established.

The next thing is to try and throw off and extract all corrupt matter and to dissolve any hardnesses which may have arisen in the system.

Should the circulation of the blood be out of order water again will most effectually clear the obstructions and restore it to its normal condition.

Kneipp, Codicil. 18

Therefore I repeat provide suitable and good nourishment for general strengthening of the system and removal of the harmful influences.

As regards the water applications the commencement must, as I have already said, be quiet and gentle so that the patient is not startled by violent treatment, therefore give him daily in the morning an upper-body-washing and in the evening a warm foot bath with ashes and salt lasting fifteen minutes but only for a few days consecutively: further, accustom him to walking in water and on wet stones. After some days pass from the upper-body-washing to the whole washing and from walking in water and on wet stones to the knee-douche which will result in bracing the system and drawing the blood from the head.

When these applications shall have brought the invalid into a better condition then give him two or three half baths in the week according to his strength and instead of the upper-body-washing give him an upper douche.

If the patient makes very good progress half-baths, full douches and whole washings may be taken in turn, for instance two half baths, two full douches and two whole washings in the week.

Exciting applications like the lightning douche should be avoided until the improvement in the invalid is considerable and even then it should be confined to one or two weekly, administered by one who is skilful and accustomed to giving them.

It must not be forgotten that water in itself acts excitingly and that strong applications may increase the exciting effect to irritation.

The dissolving and extraction of diseased matter may be accomplished by bandages as well as by washings and douches. In the use of them begin with the simplest and easiest and gradually pass on to the more strongly acting bandages.

The blood may be drawn from the head by putting wet socks on the feet which have been soaked in vinegar and water.

Obstructions in the body can be rapidly and easily removed by the short bandage which should not be used too often nor allowed to remain on too long, otherwise increased weakness rather than increased strength would ensue.

The patient must have rather a strong constitution to take two short bandages in the week and very well nourished indeed to take three in the week and even so the use of them should not continue beyond a fortnight.

Just as the short bandage from the arms to the knees acts favourably on the body so does the so called Spanish Mantle act dissolvingly and extractingly on the whole body.

Should however a mentally diseased person remain too long in the mantle it would greatly excite him and therefore it should never be kept on longer than an hour. In using the bandages proceed in the following manner: Bandage the feet as far as the ankle two or three times a week to act upon the circulation of the blood and a short bandage twice and a Spanish Mantle once in the week to dissolve and carry off impure matter.

With these weekly applications it would be well to combine a half bath or whole washing.

Some may perhaps think that vapour baths would effect the most good, but the experiments I have made with them show that this is not so and that they should be avoided because they greatly excite the patient, and should a tendency to insanity exist in him the vapour baths might help to develop it.

On the other hand I was pleased with the effect of the foot vapour bath followed quickly by a knee douche.

18*

Occasionally compresses are more beneficial even
than bandages because they act quite especially on the
abdominal organs and produce good digestion and also
because they are so easily applied to the patient.

For weakly people the compress should consist of a
four fold cloth dipped in water and vinegar and laid on
the abdomen for an hour: it may be kept on in some
cases for an hour and a half, but if so it should be
renewed at the end of three quarters of an hour. Such
a compress may be applied two or three times a week.

The lower compress twice a week and the upper
compress once will be found of great service. The former which acts bracingly and extractingly should be
kept on for half an hour and the latter for three quarters of an hour or even a whole hour.

From what has been said it may be seen that water
may be used in many ways on the patient and that the
effect on the body is to dissolve and extract corrupt
matter and to cleanse and brace the system.

As in all other diseases so also in this, the gentler
the treatment the more favourable the result: one will
never cure a hot headed passionate person by disputing
and contradicting; kind words and yielding one's own
opinion will be infinitely more effective.

The Hair and the Nails.

For many years indeed I may say as long as I have
had the power to think I have pondered much on the
signification of the hair and the nails; what end they
serve and why they were given to man.

One might think it not at all important whether a
person had little or much hair or even none at all, or
whether his nails were hard or soft.

The nails of some people are so hard that they
become brittle on the slightest occasion and this is often
the case with the hair also.

Both hair and nails are a protection to the body. Hair and nails also tell us in what condition the system is. If the hairs are thick and supple and continue to grow one may assume that the system is in a good condition.

If, however, the barber can remove the beard without difficulty we should infer that the man's power is decreasing; the softness of the hairs of the beard proclaim this: if the beard will no longer grow it is because the system grows weaker. The growth of the hair may cease or it may become thin and decayed even in young people which is a proof that the person is not strong or that an illness is pending and hindering the nourishment and bracing of the system.

It is a true saying that a child born with much hair is generally strong.

What has been said about the hair holds good also with the nails on the hands and feet. If the nails are too soft and thin and grow too slowly it is a sign that natural power is lacking. If, on the contrary, the nails on hands and feet are like good bone and if the colour is white with a rosy glimmer one judges the person to be in good health. If the nails have a bluish tinge on a leaden ground it is regarded as a sign of disease. All that we have said about hair and nails holds good with children as well as with adults.

One can tell by the nails whether the body is properly nourished or whether it be diseased.

In anaemic people the nails become blue, in consumptive people white: the hair tells quite as clearly as the nails the condition of the body. If, on the head of a little child, there is a good crop of hair one may safely infer that if no obstacle intervene the system will be strong and healthy.

For many years I looked after a farm and the creatures I loved best to rear were calves with thick close

hair for they always prospered, for those with thin delicate hair never' fully developed or became large and fat.

This is yet more striking in young horses; the
peasants say "If in the late Autumn or Winter one can
plait little pig tails on a one year old horse then you
have a healthy and strong horse."

From what has been said it is clear that hair as
well as nails were ordained by the Creator as necessary
to man.

The hair is a protection to the head and the nails
are a protection to the toes and fingers and they should
be suitably cared for like all other parts of the body.

Any one having had the misfortune to lose a nail
will know how difficult it is for the finger to do without
its protection. It is just the same with the nails on
the toes. If the hair falls off, the head loses its protection; this is serious when one thinks of the many
delicate organs existing in the brain which, deprived of
the protection of the hair, may be exposed to many
dangers of cold, heat or blows.

How difficult we should find a great deal of our
daily work if we had no nails; the fingers would be hurt
and allow poisoning or other injury to arise: thus hair
and nails are a double protection to the head, hands
and feet.

Just as each individual part of the body is subject
to disease and infirmities so also the nails and the hair.

I have seen nails on fingers and toes which were
four or five times as thick as they ought to have been,
indeed so bad were they on the toes that they were as
hard and thick as claws. I knew a gentleman who could
not cut his toe-nails at all and was compelled to have
whatever was necessary removed in small pieces. They
were so thick as to be quite a burden to him.

I advised him to take two whole washings in the
week, to walk in water one day and on the next to
wrap his feet up as high as the ankle for an hour in a

cloth dipped in decoction of pine-bark. The result of this treatment was that the growth of the nails ceased, the rottenness was got rid of and the condition of the nails became so improved that he was able to cut them with ease. Some growths on the nails are brittle like rotten wood while others are firm as bone.

I knew a seamstress whose finger-nails were so thick that she had to shave them from time to time with a knife. Undoubtedly the cause of such a disease lay in the body; it may be an ill regulated digestion or some other disturbance: this growth is more than an exudation; it is as if pitch-ran out of a fir and became hard.

One may get rid of it by general treatment for the improvement of the digestion and for extracting all diseased matter from the body.

Disease may act on the nails as part of the whole system which at the time is in a thoroughly sick and unhealthy condition. If the diseased matter is conducted out of the system then the organs will again become healthy and the evil removed. And this is brought about most effectually by a good diet, a good digestion, and healthy activity of the system, a condition that permits of no unsound matter in the nails.

Out of a good soil one expects a good crop while out of a bad soil we look for nothing flourishing. If the hairs have a good soil and are well nourished they grow unusually fast; if, on the contrary, the soil is not good or has matter in it which the hairs cannot use or which even hinders the growth of the hairs then diseases arise and the hair falls out.

Little children having thick long hair may be regarded as having good healthy hair soil: if, on the contrary, the hair is fine, soft and short we may assume that the children are weak and that their hair will not prosper. The former have beautiful hair all their life through, while the latter are always in anxiety lest in time there should not be sufficient to cover their heads.

By hair soil I mean the scalp in which the hairs are planted as plants are in a garden. If the hair does not prosper it is certain that the material must be lacking by which it should be nourished and by which it thrives. Under these circumstances the scalp will be thin, dry and poor in juices; the hairs too will be very thin, a sure sign of decay and that they do not get enough nourishment.

On the other hand with some people the hair is thick, stiff and dry so that occasionally it splits, and when this occurs it gets into such a tangle that it is almost impossible to disentangle it.

Other people again have a greasy secretion of the skin, particularly noticeable on the scalp; the consequence of this is that the hair is damp, greasy, and oily and as a rule exhales an evaporation which is not at all agreeable.

Such greasy hair naturally does not split. People who do not like split hair very often rub ointment into the scalp which produces a shiny, glossy appearance. Suitable care of the scalp and hair is very necessary and consists above all in cleanliness.

Again, the hair must have light as plants cannot thrive without it.

To secure cleanliness the head should be washed as often as once or twice a week.

I know that fashion recommends various hair washes but to my mind none is equal to that given by the Creator for cleansing, which is water. I have no objection to a harmless addition used now and again especially if it be a decoction of herbs.

For example, washing the scalp two or three times a week with pewter-grass tea cleanses the skin, prevents the closing of the pores and improves the evaporation, all of which are necessary to the growth of the hair.

Stinging Nettle also has a good effect: perhaps it is the so called Nettle acid which acts on the pores and cleanses the skin which produces the growth of the hair.

Now comes the question, **"should those who have long hair grease or oil it from time to time?"**

I think it necessary occassionally to keep the hair from splitting but it must not be done too often as it closes the pores and so produces an evil smelling evaporation.

Some people suffer from scurf on the head. **Scurf** is nothing else than the outflow from the skin of the head which does not disperse in the air but dries on the upper surface of the scalp. If a person perspires much, crusts form on the skin. The pores become closed and the air can no longer penetrate, the consequence being that the hair roots choke and the hair falls out.

This strong exudation towards the top arises generally from head work; for the more the head is exerted the more the blood presses into it and causes a strong evaporation, a closing of the pores, and a falling out of the hair.

Growing people often have scurf in the hair. This may arise from an insufficient digestion of food whereby many things get into the blood which do not properly belong to it, and so an exudation takes place by which hair scurf is formed. This should never be allowed to remain. It is best got rid of by cooking stinging nettles in half water and half vinegar and well rubbing the scalp in this decoction.

The Vinegar dissolves, and opens the pores so that the hair thrives and the scalp is cleansed. Many sharp remedies are recommended for the removal of scurf but I believe such to be harmful to the skin, the pores and the hair.

The natural treatment is always the best, the more artificial the remedy the more dangerous it is.

Each part of, the body must be carefully and separately looked after not forgetting the nails and hair.

Those who desire to be really healthy should have the necessary knowledge as to how the individual parts of the body should be cared for.

Hysteria.

It is not without reason that we call hysteria a bad and, in many respects, a dreaded disease. If we had not examples of it almost daily it would be difficult to believe that it could reduce people to such a fearful condition.

Although men do not altogether escape it, it is the women specially who suffer from it.

It is my belief that hysteria is caused by poor blood and weak nerves and that the mischief begins in the abdomen. As a rule, lack of good nourishing food is the source both of poverty of blood and loss of physical strength and in such a condition every trifle causes disturbances in the system and the whole body becomes gradually weaker.

This shows itself either by very great debility or by great excitability; if the surroundings are favorable the former will prevail if not the latter will gain power.

As hysterical people usually look well, people do not realize that they are ill and put down their complaints to a highly excited imagination, forgetting that no human imagination unaided could produce such a condition.

I knew a very good girl with excellent parents of simple unpretending way of living. She was an industrious school girl and the joy of her parents. As she grew older she displayed special peculiarities towards her people being either very aggrieved or very excited.

When they felt it necessary to oppose her or dissuade her from any step she maintained her own opinion with great violence, and at last things became so bad that she had but two moods the one in which she was intensely excited, the other in which she was utterly depressed and powerless.

Indeed so great were the disturbances in her system that she was frequently unconscious and so far removed from a normal condition as to do quite unheard-of things.

In her moments of exaltation she wrote letters in the purest German, models as to form and matter; even perfectly new words came out in them, so that students asked how anything like it could be possible.

She made observations and passed judgment in such a way that we all wondered how a simple girl could have such universal knowledge and keen insight.

She wrote the most beautiful little poems without ever having heard of poetic metre. At other times she felt so deadly weak that she could scarcely speak nor did she trouble about anything that went on about her. Nothing could be a greater contrast to her former mood.

She was just as extraordinary in her choice of food; she seldom took that which had formerly pleased her and what she preferred on one day disgusted her the next.

In this condition she remained about ten years sometimes better sometimes worse. Although we know that hysteria is no mere imaginary disease yet we must confess that a diseased imagination exercises a great effect on such a person and that it is impossible to dissuade her from anything she has once decided upon.

If a person so suffering thinks she can walk, she walks: if she believes she cannot no one can make her although she may be able to walk quite well.

If hysterical people have faith in their doctor they will take his remedies which always take effect just on account of this confidence, but if the most skilful doctor

in the world prescribed for them they would not take
his remedies neither would they be of any use if they
did, providing they had no confidence in him.

One can say with truth that such invalids are doubly
miserable because they are really ill, though not in the
way they themselves think and because they suffer from
fixed ideas whereby their condition gradually grows
worse.

People suffering such an utter wreck of their nervous system often show symptoms of disease in curious
and diverse ways.

One patient will laugh at every trifle so as to get
cramp from the action; another will cry and become
troubled and comfortless without knowing why.

Others are unstable and irresolute while others again
carry themselves to all appearance with great dignity
and are insulted at every trifle and deeply so.

These conditions often continue until they end in
convulsions which last hours and even days.

In spite of individual organs being healthy there
still prevail in the body these diseased, disturbing influences.

It not infrequently happens that, with hysteria, paralysis appears, it may be of the vocal cords so that
the invalid cannot produce a loud word or in the hands
and feet or other parts of the body so that it is thought
the invalid must have suffered from a paralytic stroke.

I know cases in which the extremities were completely paralysed; if one uses the proper remedy this
may sometimes be cured at once.

People have come to me who for months or even
a year have not been able to produce an audible sound,
yet on the application of the first thigh douche their
speech returned.

One must not however think that complete cure
follows on a single application; the cause is merely
removed which induced the paralysis.

Thus such a person presented herself to me having entirely lost her voice. She came at two o'clock, had an upper douche at three o'clock and shortly after regained her voice. It must not be assumed that regular paralysis existed in this case but only a passing disturbance in those nerves which set the organ in motion.

Therefore one must take special measures as regards the paralysis but at the same time operate on the whole body. As this last gets into order the individual defects will be removed.

Some people attribute these diseased conditions to Celibacy and advise the invalids to marry telling them that if they do so they will get out of their sad state. All I can say is that innumerable cases have shown me quite the contrary.

As a rule hysterical people lead very good lives; of course there may be exceptions. To many of these I have heard the advice given "Marry and you will be free from your affliction". Many have followed this advice and after a few years find that not only they have not lost it but that it is present in six-fold strength; a clear proof that hysteria by no means has its origin in Celibacy.

Now comes the question "How can such a condition be cured?"

I consider three things to be specially necessary and important; first the invalids must have perfect faith and confidence both in the medicine prescribed and in the Doctor, for this is just a case in which no pressure can be used; secondly, **bracing** the system must be tried and thirdly a diet must be chosen which will be sufficiently nourishing to make fresh blood and give to the body increased health and strength.

The food should be easily digestible and full of nourishment while all that excites the palate should be avoided, so should all alcoholic beverages, as they increase the bad symptoms rather than improve them.

Water should only be used in the mildest way at first, gradually a very gentle increase may be made.

At the same time take good notice whether the invalid has an aversion or dread of this or that remedy and if so remove it and try as much as possible to follow that which she prefers or likes unless it be something injurious.

The patient should walk on wet stones for one or two minutes daily for several days and wash the whole upper part of the body every morning with water and Vinegar.

By walking on the wet stones, the patient is braced and the blood is conducted to the extremities where generally the greatest poverty of blood exists.

By the washing an increased natural warmth sets in, the upper part of the body is braced and the transpiration, which in such invalids is mostly very weak, is brought into a proper state.

After some days the patient may in addition to these applications wash the lower part of the body with vinegar and water which will bring more life and activity to the whole system; then on one day a whole washing from bed, on the next a knee douche and a walk in water, on the third day a thigh douche continuing in the mean time the whole washing.

If these applications are persevered in for some time and followed up with a rapid half bath on one day and on the next a whole washing alternately with an upper washing, the system will become thoroughly braced and the health much improved.

Soon after one has commenced the washings of the Abdomen a four fold cloth should be dipped in hay-flower-water and laid warm on the Abdomen three or four times a week; this decoction operates very favourably on the Abdominal organs.

One may try to give a lower compress one day and an upper compress on another day about three times in the week; they need only be kept on half an hour to ensure great benefit to the patient.

One can operate internally as well but only in the simplest way; a tea of oat-straw and honey is not only very strengthening but dissolves the diseased matter which has collected in a weakly system. A table-spoonful should be taken every two hours for one or two weeks.

A similar tea may be made of Oak bark or acorns mixed with sage or ribwort. The former strengthens the system and the latter promotes good digestion. Those invalids who prefer bitter things would do well to take tea of centaury, bitter clover and wormwood. Omit the last ingredient if the patient has an objection to it.

A tea made of St. John's Wort, common Yarrow and camomile may be given as a change but the tea which operates most searchingly is that made of Angelica and Tormentilla.

A tea which both strengthens the system and prevents attacks is made of Valerian root and rue, an equal quantity of the herbs are boiled together with honey and the tea thus prepared is taken in small portions every two hours (a teaspoonful) or a tea may be made of each herb separately and taken alternately; in many cases this is more advisable because this disease demands a variety in the remedies.

If the patients have a great preference for any particular remedy it rarely lasts longer than a week therefore change the tea often and even that which disgusted them at one time will be acceptable to them later.

With some of the invalids tinctures of the herbs already mentioned operate very well though some cannot touch alcohol in any form; in this case tea is the only remedy.

The herbs which make the best tinctures are Tormentilla, Angelica root, Juniper berries, Valerian and

Wormwood each taken separately; six to ten drops three times a day on sugar or in a spoonful of water will suffice.

Whooping Cough.

Whooping cough is one of the most malignant and dangerous of childish diseases.

Children suffering from it are in constant danger of suffocation from the violent attacks of coughing.

It is also infectious; if once a child in the house has it every other child in the place follows the example.

At first whooping cough begins with a little irritating cough and fever which condition gradually increases and developes into violent attacks.

If we observe a child in one of the fearful paroxysms of coughing we notice a great access of blood to the head, a red face and bluish red lips followed often by severe bleedings of the nose and a bloody foamy slime from the mouth. These paroxysms of coughing may repeat themselves thirty or forty times a day. According to the violence of the attacks one can judge of the improvement or reverse of the condition.

Danger of infection has to be guarded against and therefore the healthy must be completely separated from the sick.

As far as possible the greatest quiet must be secured as the slightest agitation or noise gives rise to these fearful paroxysms.

One cannot do much in these cases with compresses or bandages, for the moment the attack comes on the child tears them off and the last state of things is worse than the first.

Then, what can one do for such a patient? I have found that the best thing is to take the child out of its warm bed once or twice a day and dip it for a second in cold water and put it back again into the warm bed. The first day the child will resist naturally but when the application has been once made it will bear its repetition willingly. Giving medicines is of little use in whooping cough, rather give nourishing food which is also a remedy for cramp.

A very suitable thing is Fennel or Anise tea, or fennel cooked with anise in milk. For the rest a good strengthening broth should be given, for the child cannot well manage any other sort of food. This treatment is the best and easiest for the disease. By dipping in cold water the system is freed from the corrupt matter which causes the irritation.

Whooping cough frequently leaves behind little troubles affecting the lungs or some other part of the human system, therefore care must not cease with the cure of the cough. It is necessary for the patient to take simple water applications and to live for a time in the Country where he or she may have fresh air and by these means to strengthen the body and free it from all diseased matter.

"Housemaid's Knee".

Housemaid's knee is a very troublesome evil produced by much kneeling.

It begins with a small swelling which gradually increases and frequently passes on into abscesses.

As soon as one sees that a swelling is forming on the knee it should be taken in hand at once. Many things have been tried but I have found a plaster of Fenu-Greek by far the best and most effective remedy.

Take a warm pap or pulp of Fenu Greek and lay it on the knee and leave it there for an hour. By this

means the swelling is reduced. If, however, the housemaid's knee is in its very early stage a bandage of
Fuller's Earth water will suffice. In any case take a knee
douche daily.

~~~~~~~~~~~

The Itch (itching maggot).

(Sarcoptes hominis seu scabia.)

The itching maggot is a little creature and has,
according to microscopical examination, a broad longish-
round body about a third of a millimetre long covered
with hair and bristles. It bores into the skin three or
four millimetres deep in an oblique direction and there
lays the foundation of a loathsome disease which is called
the itch (scabies).

It is a well known disease most frequently to be
found where great uncleanness prevails in house, clothes
and beds; one is not far wrong in calling the itch
vermin.

A person attacked by this has no rest night nor
day yet his condition is worse at night when the whole
body is equally warm.

The first symptoms of the disease show themselves
as a rule between the fingers in tiny blisters which
gradually spread over the whole body. Out of these
little blisters comes from time to time a pus-like fluid.

The mode of treating this disease is varied; some
prefer the English*) itch-salve, a mixture of Sup. virid.
Sulf. subl. etc.

Others rub the whole body with so called Green
soap and some people content themselves with internal
remedies.

---

*) N o t e. Another salve is made of equal parts of sulphur
ointment and nitrate of mercury ointment rubbed together.

In my youth I used to hear of this disease and that they tried to get rid of it by a salve made of brandy, flowers of sulphur and grease.

With this warm salve the patient was rubbed and the result was a loathsome smell throughout the house; other similar salves were also used. It seemed sometimes as though this rubbing freed the body of the evil and restored health to it. Yet this in most cases was a delusion, for the invalid began to suffer so many inconveniences that one could not but think the Vermin were carrying on their work of destruction internally.

Thirty years ago I tried to cure the itch with warm hay-flower-baths of 26^0 to 28^0 lasting a quarter of an hour with cold ablutions following immediately.

The application was repeated three or four times, fresh linen was put on after the process and the bed equally provided with fresh linen. In my opinion this treatment is the best.

If the tiny creatures have been long at work in the body one must act first upon the organs of the stomach and the digestion and the most suitable means for attacking these is wormwood tea or wormwood and sage combined of which a spoonful should be taken every day.

In time I made experiments with cold water alone, for I asked myself "Why should not the same result be reached by cold water?"

Warm water has one advantage viz. that it brings the Vermin quicker to the surface but as warm baths weaken the system they should not be taken often.

For many years I have almost entirely given up warm baths and have turned more and more to cold water.

It certainly takes longer to hunt these creatures out with cold water, yet on the other hand its effect on the whole body is of a strengthening and bracing character.

I recommend every night from bed a whole washing, beside two or three half baths and a full douche in the week.

19*

During the cure the body and bed linen must be constantly changed so that the vermin cannot spread.

The food should as usual be simple and strengthening avoiding strictly vinegar and spices.

If any one should ask whether cold or warm treatment is the better, I answer that the warm gets rid of the vermin more quickly but the cold is much more strengthening for the body. To dispose of doubts I advise that both methods should be tried; take a warm bath daily for two or three days with a cold ablution immediately after each. Frequent change of linen is absolutely necessary.

When the extraction has succeeded look after the strengthening of the body and take some half baths in the week alternately with upper and full douches.

Cancer (Carcinomus).

Cancer is a disease by no means rare but I know of none more dreaded.

It is possible to get this disease by imagination, cowardice and fear.

The question is frequently put to us, What is Cancer, and how does it arise?

In my opinion it is simply an illness in the blood.

As the blood can be made free of impure matter so on the other hand may it be exceedingly corrupt from the amount of foul matter within it.

Blood is formed from nutriment and it may be assumed that the more suitable the food the better the blood.

He who consumes unwholesome food cannot be surprised if unwholesome matter gets into his blood. There are people who have bad blood all their lives long while others, on the contrary, have always good healthy blood.

When unwholesome ingredients get into the blood the system desires to eject them and in doing so causes sores to appear, this is her way of getting rid of the harmful matter. Just think of scarlet fever, measles, pocks, and similar diseases; these eruptions are all more or less poisonous and may not only harm the system but ruin it, therefore it throws them off.

Or look at a Carbuncle or Abscess, these are nothing more nor less than ejections of the blood.

Among these ejections effected by Nature which are all more or less inflammatory and poisonous, Cancer is certainly the worst and most malignant, seeing that it is incurable when once it has got the upper hand.

Cancer therefore arises from impure blood in which so many foul things exist that the good blood succumbs and the foul alone is left for the body's nutriment. Of course this is not in a condition to keep the body sound and as the foul matter gains ground an outbreak of it occurs somewhere on the body, assisted it may be by some obstruction in the blood.

The younger the person the less is this disease to be dreaded because the blood has not had time to absorb so much corrupt stuff, the system is still active and the circulation of the blood is better regulated than with older people.

If, however, the activity of the bodily organs decreases. the circulation of the blood becomes slower and more imperfect, the obstructions in the blood more serious, in short, if man is past the prime of life the sickly stuff gains power and conquers the system, and the body's capacity for resisting the diseased germs grows less.

This disease may be caused by a natural obstruction in the blood or by an obstruction caused by a blow, or knock or pressure.

If cancer has entered the system and the blood is spoilt by diseased matter it may appear in several places,

passing from the parts first affected to other parts of
the body especially where the obstructions have been of
longer duration.

Again, the question presses for answer "Is Cancer
curable?"

It is true that by my treatment many people, who
were considered by Doctors to have Cancer, have been
healed; yet if this fearful disease is spreading havoc in
the body cure is no longer possible, for the blood is too
corrupt and it is too late to replace it by other and
healthy blood.

If, however, the system is in a tolerably good condition and the age of the sufferer not too advanced quite
unexpected results may be obtained by Water, which
has an extraordinary effect on the blood and juices en-
abling the system to eject all foul matter.

Herbs also are excellent in their power of renewing
and improving the blood and juices and I repeat what
I have said in another place, viz., that herbs do more
even than food or medicine to improve the quality of
the blood and nourish the system.

Cancer may appear in various parts of the body;
we will describe one or more of these.

Cancer in the Tongue.

Cancer in the tongue is not very common.

It may arise from bad teeth, from wounds or sores
on the tongue or from the habit of biting one's tongue.

It is therefore advisable to have the teeth frequently
examined to see that no tooth exists which would by
any possibility produce this evil.

If the mischief has begun, gargle every hour with
pewter grass tea which will cleanse both tongue and
teeth.

Cancer may arise from a trifle but it increases rapidly like mice in a field. Of the miseries caused by

Cancer in the tongue there are examples and to spare. Like all other Cancerous diseases that in the tongue is very infectious.

Cancer in the Stomach.

Formerly Cancer in the Stomach was called "a hardening" in order, as I think, to prevent the feeling of dread and anxiety which the very name of Cancer produces.

As I look at the matter it happens in the stomach like warts on the hand, as if moles threw or built up a hill. The stomach having much blood it frequently clots on the inside of the gastric mucous-membrane producing little knots in form. like warts on the skin. Nourished by the blood they grow larger in size and in number, are firm and hard and do not break.

In this way many such growths may form near one another getting ever larger and maintaining their firmness till they become big swellings The larger the swelling the more numerous the knots and the more **serious** the troubles in the stomach.

At first the matter seems unimportant and as a good deal of mucus is excreted it is thought to be Gastric-Catarrh.

If this supposed Gastric Catarrh lasts very long, say for a month, many disturbances appear, food is rejected and pains increase till they become almost unbearable. The complexion alters, the face becomes blotched, tired and faded and one has the impression of being in for a bad illness. As the colour and features of the face alter so the patient gets thinner even though for the time he is able to make nourishment.

Not only is the blood no longer pure but it is not properly distributed thereby depriving the patient of much of his natural warmth. As this decreases the condition of the man becomes worse.

Suppose the hardness and formation of knots have existed for a year or longer the knots begin to break

in various places, just as swellings on the outer surface
of the skin burst. This naturally causes the filth which
has escaped to corrupt the healthy parts of the system
and completely destroy the nutritive ingredients so that
the whole man is diseased.

By this discharge the gastric mucous-membrane
becomes more diseased, the swellings spread and become
more dangerous and the veins being overcharged it only
needs one of them to burst to cause the man's or woman's death.

When this evil developes in a man he complains of
burning and stinging in the stomach and of great inconvenience. He cannot take much food: he has an aversion to those articles of food he formerly liked and
prefers those which he would not even look at in his
healthy days showing how out of order his system has
become. Naturally his strength decreases and his mental
condition is variable and uncertain.

After partaking of food the stomach becomes distended by gases, he suffers from a feeling of pressure
and an ever increasing thirst, but he persists in thinking
himself better than he really is and, like the consumptive patient, refuses to see death when it stands close
to the door.

As the disease makes progress frequent vomiting
occurs; at first the vomit is like white slime, later the
colour gets so much darker that the patient often says
it is Coffee-grounds.*)

The Action of the bowels is generally very hard
and increases till medicines are no longer of use and the
patient has to resort to injections.

The Urine passes sparingly and is full of sediment.

---

*) N o t e. The conviction will have gained ground that if
Cancer forms at the entrance to the Stomach vomiting follows
the eating rather quickly; the nearer, however, it is to the exit
the longer will be the time between the meal and the vomiting.

In Cancer of the Stomach as in other diseases there are three stages; in the first the patient thinks his stomach weak but suspects nothing malignant; in the second stage great inconvenience and vomiting occur, the Cancerous knots are perceptible, and emaciation makes progress. The third stage is that of absorption and the end rapidly approaches.

If impure and ill-regulated circulation of the blood be answerable for this disease then the cure must be attempted by means of improving the blood as in all diseases which have their origin in the blood.

If one is not quite sure whether it is gastric catarrh, gastric tumour or Cancer it is in any case good · to operate both internally and externally. I recommend strongly for the first, wormwood, tormentilla, Angelicaroot and pewter-grass. A weak tea of these herbs may be taken daily; a cup full in two or three portions, supposing the disease to have been grappled with on its first appearance; if, however, one is of opinion that it is not gastric catarrh, but either gastric tumour or Cancer, then take a spoonful of the tea every hour.

Other herbs however may be used and indeed those are best which one uses externally for healing wounds and sores, for if they operate in cleansing and healing externally I see no reason why they should not exercise the same power within.

We know, for example, that tea of Chickweed used as a compress cleanses even old sores so well that we are amazed that so insignificant a herb can bring about so great an effect. If then this herb cleanses and heals external sores would it not have an equally good effect on internal sores and growths?

Again Oak-bark operates so excellently on external sores that it cauterises the foul matter and heals those which are most loathsome.

Angelica is good for wounds caused by knocks or blows and Ribwort for fresh wounds and old injuries.

If then such herbal juices penetrate internally why should they not also procure cleanliness, improvement and cure always provided that the mischief has not gone too far ?

These herbs are all good even when used separately and when two or three are used together, they are even more effective for cleansing and cauterising; or the herbs may be used with advantage alternately.

We shall be able to accomplish more by water externally in this disease than internally yet its application must be arranged according to the stage it has reached.

I have had many sufferers here in Wörishofen whose diagnosis ran "Either Gastric tumour or Gastric Cancer". Their bodies were in a tolerably good condition and a fair amount of natural power and warmth still existed.

In all these cases I found that the half bath cleansed, warmed, dissolved bad matter, and strengthened the whole abdomen and that the upper douche did the same for the upper part of the body especially the back; and that the full douche conduced to general strength and activity so that the sick people acquired more warmth and a better digestion and the blood became purer and flowed more freely, because of the impurities which had been ejected and thrown out.

As one gets to know the state of the patient more certainly compresses may be laid on the stomach and abdomen if he complains of difficult action of the bowels or oppresion of the abdomen which depends on Gases; they should not be applied more than twice in the week.

Bind a four or six fold cloth which has been dipped in warm hay-flower-water on the Abdomen in the region of the stomach and leave it there for an hour and a half if you wish to dissolve matter; if on the other hand much gas exists dip the cloth in cold water and Vinegar and renew it after three quarters of an hour.

If clear symptoms of advanced disease show themselves in the patient as, for instance, emaciation, loss of appetite, sunken features etc. then the treatment both internally and externally must be more gentle. He may be able still to bear the half bath but I would rather that he should wash his whole body night and morning and take either a knee or thigh douche during the day. If after these applications tolerable power still exists the back douche and half bath may be taken, and supposing that with this treatment the patient's condition improves there is a prospect of cure or at least a lengthening of life.

At this stage never give a short bandage but instead a compress on the Abdomen for three quarters of an hour; if necessary it may be kept on an hour and a half but in that case it must be renewed at the end of the three quarters of an hour to avoid the great heat which might arise and cause the patient trouble.

For some few days the Compress may be taken daily for three quarters of an hour but subsequently twice a week will suffice.

If the Cancerous disease has advanced so far that a general decrease of power ensues, vomiting becomes frequent and life is threatened, then only the very weakest applications should be taken just sufficient to subdue the great heat, freshen the body and strengthen it as far as possible. These will subdue the pain and enable the patient to bear his sorrow more easily.

At this stage it is possible to lay on the Abdomen a simple small cloth which has been dipped in water, or water and Vinegar, or in hay-flower water, according to which acts the best, but whichever it be it should be renewed as soon as it gets hot. If possible let the sufferer have an upper or whole washing as he will find greater help from these than in taking drugs because by the Washings the skin is made to perspire and becomes refreshed. One may use water alone for these Ablutions or mix it with vinegar or good brandy.

As regards nutriment the patient must himself choose that which he can best digest, always avoiding acid or highly seasoned food as well as that which is known to be indigestible.

Of the herb-teas already named let him take the one which suits him best but only in small quantities and very weak.

In place of tea he may use tinctures if he can take the alcohol they contain, some people are the better for them while others find them most unsuitable; Never let more than from six to ten drops be taken mixed with water.

At the commencement of the disease and even in the second stage powders of herbs may be used with great advantage; Angelica root powder cleanses the diseased internal parts of impure juices; Wormwood powder is equally good and Oak-bark powder has been used with great success.

Seeing how very fearful a disease gastric Cancer is one should guard against it from earliest youth by taking good nutritious food and avoiding everything that injures the stomach.

He who lives reasonably and practises moderation will be able to defend himself from this and many other diseases.

Cancer in the Rectum.

As there is Gastric Cancer so also is there Cancer of the Rectum. As blood obstructions cause growths in the stomach so here in the Rectum they produce growths and tumours.

Swellings which at first are small and unimportant become gradually larger and cause a contraction of the Rectum which in its turn naturally creates a difficulty in the action of the bowels and thus the situation becomes ever graver.

If these swellings get old and begin to decompose they break and we have the complete cancerous sore.

The discharge from these internal sores carries the mischief further resulting often in severe hemorrhage and naturally an intense weakening of the system.

As the features of the face fall in and the complexion gets worse so the strength vanishes and the appetite goes, and owing to the great poverty of blood together with the blood poisoning life comes to an end.

Sores dangerous or less dangerous may be cured or at least improved by decoctions of herbs.

Seeing that Cancer in the Rectum developes very gradually into sores which are not at first malignant a good deal may be done by internal and external operation.

If decoctions of herbs can be conveyed into the Rectum by a syringe the sores will heal if the disease has not gone too far. I recommend as the best remedy a decoction of oak-bark and pewter-grass; Fenu-Greek is also most useful.

For Cancer of the Rectum, Water applications not only strengthen the body but diminish the heat and extract the bad matter. Those which act best are the thigh and back douches; half baths and compresses alternately twice in the week.

To operate equally on the body administer also one or two upper washings; avoid, however, too many applications as the great object is to strengthen and brace the system.

If the invalid is so weak that he cannot take douches let him have instead one or two washings.

Even if the mischief lies ostensibly in the Rectum yet it has its foundation in the blood; this being so we must improve it by the administration of teas made of blood-cleansing herbs of which a cupful daily in two or three portions should be taken.

Cancer in the Breast.

Cancer in the breast appears rather often with women but rarely with men, of course there are individual cases. Just as other kinds of Cancer this also has its origin in impure blood, though it may now and again be caused by some external pressure or blow. Several people suffering with Cancer in the breast have told me that formerly they had certainly carried articles by which special pressure was caused on the breast and produced pains but that it was not till some time after that they observed the formation of a tiny swelling which, however, was not specially painful unless pressure from without touched it — indeed many of the people had had such a lump for seven or eight years or even longer. As they saw no material increase they thought the lump harmless and did not come for help until the mischief became very noticeable and then it was often too late.

Still by means of water and herbs many such knots and swellings were completely dissolved and healed and it is difficult to say whether these knots were already cancerous or whether they would have become so in time.

As long as the lump is only on the breast, and glands in the vicinity do not become hard, one may assume that healing is possible and I must say I have had the greatest success in such cases. If, however, the glands under the shoulder harden from that point to the breast then we may conclude that the mischief is in an advanced condition.

In young people of from twenty to thirty years of age such formation of lumps is not as a rule dangerous; on the other hand they are very serious where they occur in people between forty or fifty years of age and are difficult of cure.

The cure should be attempted in a double form; first by improving the blood and secondly by dissolving and extracting the bad matter and strengthening the whole system.

The external application of water by means of bracing, extracting and dissolving brings the circulation of the blood into order and causes a general healthy activity of the body.

The Application should consist of two half baths, two thigh and upper douches, one back douche and a whole washing in the week. Even if the patient has her full strength she should not take more than two applications daily, one of which may be weak and the other strong. In this way the body is acted on generally while the part affected receives special care by means of compresses one of the best being of pewter-grass. A soft small cloth is dipped in pewter-grass water, well wrung out and laid over the whole breast; this is continued the whole day, the cloth being redipped every two hours. If the swelling is large, as big as a fist for example, pot cheese is a matchless remedy.

The pot cheese is well beaten up, spread over the breast and left there until it is quite dry when it should be renewed. This should be given in turn with pewter-grass, or loam water compress twice a week.

The loam water compress collects the foul matter and absorbs it, it also takes away all superfluous heat and is a good alternative to pewter-grass though it may not be quite so powerful a dissolvent.

If pewter-grass be boiled with tormentilla the effect is even better. To make loam very effective it should be thoroughly dried and powdered and mixed with pewter-grass or tormentilla water. If used as a salve it must be renewed when it dries or causes heat.

If the Cancerous swelling is still unbroken it is possible that these compresses and salve may diminish the size or even cause it to disappear.

Cancer of the Lips.

Cancer of the lips is not more rare than Cancer of the tongue and its curability depends upon the strength of the patient's constitution. It is like other Cancerous diseases difficult to cure and although an early removal of it by operation is the best thing to do yet one can never be sure it will not return. Pot cheese is a most important and valuable remedy. If it can be bound on it extracts the heat from the inflamed spot and absorbs the foul matter. I have found it of service not to use one remedy only but several, one after the other; for instance, Aloes, powdered and strewn on the wound, absorb the impurity, cauterise and heal, while Alum cleanses and heals.

Aloes and Alum mixed and strewn on the bad place will greatly help.

Supposing, however, that no open wound exists but only a cancerous lump then I prefer pot cheese to all other remedies. I have had very good results with Cancerous-Ulcers by using Compresses made of hemp-agrimony-tea and by washing the wounds out with it.

Cancer does not confine itself to those parts of the body specially mentioned in this chapter but may appear anywhere.

For years one may have had a scab or crust on the body but all at once its character changes, the edges of the crust harden till at length there appears a malignant cancerous Ulcer.

Such Ulcers come often in the face in the vicinity of the eyes, on the cheeks, near the ear, on the neck and in various other parts. The treatment should be twofold; first that of the whole body because in such a condition the blood must be very unhealthy, and secondly on the diseased part itself so as to heal it by dissolving and cauterising.

To soften the swellings I make use of several remedies, which I take in turn. For example, I order beaten-up pot cheese to be laid on and kept there for four or five days and to renew it as often as it gets dry; the effect of this is to dissolve and absorb.

Later, I order a decoction of pewter-grass to be applied and renewed every two hours. I then order the use of loam or fuller's earth which invariably does good service; it is first dried, then powdered and made into a fine salve and laid on.

Its good effect is increased when mixed with tormentilla and pewter-grass waters.

If the Ulcers are wide spread I use Fenu-Greek which both alleviates and dissolves or I rub in on one or two places near the Ulcers a drop of "Malefiz" oil. This causes a strong eruption which diminishes the malignity of the Ulcer and not infrequently heals it.

From all that I have said it will be seen that Cancer is a very great evil and often incurable, therefore it should be every one's concern to protect themselves specially from this evil.

I know of only one means by which this may be effected and that is a gentle water-cure by which the whole organism is operated on by cleansing and keeping it clean. This does not allow the bad conditions to put in an appearance.

People who suffer from Cancer are generally avoided because of the fear of contagion. This danger does exist but not in so great a degree as people think, but every one should, as much as possible, keep away from the smell and dirt of the Ulcer and by constant fresh air keep them out of the blood.

Above all the chief necessity is great cleanliness both in the invalid and in his surroundings and second to this is suitable nourishment.

Liver Diseases.

Jaundice.

The human body possesses an innumerable number of glands both small and large on the skin, under the skin, in the mucous membrane, as well as in the interior of the body. The spittle or saliva is neither more nor less than a discharge from the glands in the palate.

Each gland lives and is maintained by the blood; every gland has a perfectly settled purpose and ejects matter peculiar to it. Among all glands in the human body the liver is the largest and therefore the most important. It has the task of preparing the gall which is nothing else than its secretion and serves to effect the last process of, and the most difficult, digestion of food.

It happens as in a meeting of the waters where the water of several different springs is led into one common pipe. This common Canal in which all secretion from the liver comes is generally called the Gall-duct. This conducts the secretion into the gall bladder which is the common receptacle for the Gall.

From the Gall bladder another Canal conducts the Gall into the Duodenum and there it is mixed with the nutritive ingredients which it digests with more power than spittle and Gastric juice would be able to do.

This is the function of the liver.

Beautiful and harmonious as this arrangement is it is possible for irregularities to occur in this wonderful machinery.

A water conduit may be disturbed if unsuitable matter gets into the pipes; so in a similar way can disturbances set in in the gall-duct which in their turn cause obstructions so that nothing can proceed in the proper way. If the obstruction is not removed the gall again presses backwards; it retreats into the blood, discolours it as well as the mucous membrane and the outer skin, and thus **Jaundice** sets in.

As long as these disturbances continue and the Gall presses back into the blood jaundice will continue; but so soon as they are removed the system recovers.

These disturbances may have various origins. A violent excitement may cause a block in the gall and be the means of its passing into the blood, so may a great annoyance, or fright, or blow, or pressure.

A retreat of the gall into the blood may be caused by internal pressure on the Canal, and by inflammation specially in the stomach and duodenum.

This kind of jaundice is not at first dangerous but becomes so if the disturbances are not removed; the colour of the skin alters, it becomes brownish-yellow and the condition is known as **"black sickness"** which is neither more nor less than an entire poisoning of the blood caused by the Gall pouring itself constantly into it.

If the disturbances cannot be removed the blood poisoning ends in **death.**

The invalid may not himself notice the rapid progress of the disease, although he may see that his complexion is changed, that he gets weaker, that he often feels depressed, that his nights are restless and that his appetite fails.

In this condition the sufferer complains of itching of the skin and strong biting sensation all over the body, the urine gets a dark colour, the action of the bowels is hard, sticky and of a very bad smell.

The popular ideas of this disease and its cure were formerly peculiar. I will give some examples.

A peasant had jaundice; he was advised to bind a live crab on to his back and let it die there. Another, suffering from this disease was counselled to tie some cockchafers on to his back and there let them struggle until they were dead when he would find himself free of the jaundice.

20*

Another was ordered to bind a raw piece of meat on his shoulders and after some time to take it off and hang it up in his room so that he might be free of the jaundice.

These and many other absurd things were advised and used.

That the patient was freed from the jaundice may have been true but he would have recovered just as fast without these peculiar remedies. For if the disturbances are limited, they remove themselves from the system; if they are complicated and beyond control all these remedies could not remove them and it is folly to believe in such absurdities.

Jaundice is cured when the disturbances in the Gall are removed and all the organs once again in order; and this will only happen when the obstructions are dissolved and the current of the blood follows its proper course.

Above all then it is necessary to extract and get rid of the foul stuff and then to bring the blood and Gall into proper working order.

If one operates generally on the Abdomen so as to extract and dissolve obstructions, the obstructions in the Gall-duct will also be removed and if a quicker and stronger circulation of the blood be induced then the disturbances in the liver will also disappear.

The best way of arriving at this improvement is by applying compresses of hay-flower decoction or of Oat-straw water as warm as possible to the Abdomen. These warm compresses act powerfully in dissolving and extracting and in removing the internal disturbances.

Soak a four or six fold cloth in scalding hay-flower water or water in which oat-straw has been boiled for twenty minutes and lay it on thoroughly wet. Two such compresses should be given daily and kept on for one hour if the person be weak and two hours if the patient be strong; in the latter case it must be renewed at the end of an hour. Over this Compress a woollen cloth of two or three folds should be laid as it promotes the extraction.

The invalid may also take one or two water applications daily according to his strength; the most suitable will be the thigh and back douche, half-bath and full douche; for four or five days he might take a half bath, an upper washing and a compress; then for the next three days a back douche, thigh douche and compress and lastly for another three days a full douche and thigh douche. These I should prescribe for a strong person; but for a weak patient I should prescribe a whole washing and one or two compresses daily according to his strength. If he is not too weak one application daily may be possible for him, for example a half bath one day, a thigh douche the next, a back douche the third, a full douche the fourth and so on.

For many people it would be sufficient if they took a half bath each day if by it they could get sufficient warmth; every second day an upper body washing should be taken otherwise the treatment would be one-sided.

Internally, the stomach should be specially treated so that the chyme may be improved as much as possible and easily digested so that good stuff comes into the blood and makes it healthy once again.

Equal care must be taken to dissolve and extract the foul stuffs. I have found the greatest benefit from the use of a tea made of wormwood, pewter-grass and juniper-berries of which I have given a table spoonful every hour. A second tea scarcely inferior to this is made of sage, Angelica-root and tormentilla which acts on the stomach and food; a third tea is made of oak bark, juniper-berries and pewter-grass and six or even as many as ten juniper-berries eaten daily I strongly recommend.

The yellow colour remains with jaundiced people naturally for some time and it is folly to think one can suddenly be free of it; for the gall which has been pouring into the blood can only be removed when new and better blood takes the place of the old. For this, time is wanted, in the meanwhile recovery is not retarded and the powers increase by degrees till finally full health returns.

This cure is indisputably the simplest and the surest.

Inflammation of the Liver.

Just as inflammation may attack various parts of the human system so it may attack the liver.

Wherever inflammation forms in the liver there the blood streams in increased mass; the liver thus gets overfilled with blood, swells up and may possibly undergo a severe enlargement.

If the inflammation or the mischief resulting from it is not removed the liver becomes diseased and gets into the condition known as **hardening** of the liver which deprives it of its usefulness and causes death.

Such an inflammation may last only for a short time or it may go on for a year; it depends upon the method of attack whether it is general or whether it is on one individual part from whence it spreads to the whole liver, finally rendering it useless.

Even if only a small part of the liver is inflamed it affects the whole body and therefore in trying to cure it the whole organism must be acted upon; the result will be more rapidly effected than if only the inflamed part were operated upon.

In this as in other diseases the same means must be employed to bring the body back to its normal healthy condition viz., Dissolving and Extracting.

To bring this about a **wet shirt**, a **Spanish Mantle** or a **short bandage** daily are each and all admirable.

For very weak people one would rather employ a whole washing or half bath with upper washing once or twice a day.

One operates best on the inflamed part of the liver by lower and upper compresses supposing the patient to be strong but if he is weak a simple compress once or twice a day keeping it on for one or two hours if he can bear it so long.

The Compress during the inflammation is best made with Vinegar and Water and re-dipped every hour; for complete healing, after the inflammation, compresses of hay-flower, oatstraw, and pine bark decoctions answer best.

Internally the teas given above for jaundice can quite well be used as all operate towards dissolving, extracting and strengthening.

As all inflammations leave behind more or less pre-judicial results for a longer or shorter time, so it is with inflammation of the liver, therefore the treatment must be continued for some time even if the chief cure is over.

During the week one should take a half bath or two whole washings once or twice and a compress on the Abdomen twice.

For the cure of any disease especially that of the liver it is necessary to live a reasonable and sober life.

A foolish frivolous way of living, together with an undue use of alcoholic beverages, is often the cause of liver disease. We are apt to think much too lightly of that which has become a habit; for instance, I have heard many say "I do not drink much beer or wine, at the most four or five glasses and I am never drunk"! And notwithstanding the four or five glasses daily he believes himself to lead an eminently sober life, and yet it is just this way of living which exercises the most prejudicial effect on the liver and other organs. Besides this a great deal depends on a man's or woman's profession and on the character and health of parents and remoter ancestors. In some cases even the smallest quantity of alcoholic drink produces the worst results.

Therefore I beg all who read this to guard against an irregular way of life and to test both nourishment and beverages to discover whether they are good or harmful.

Cancer of the Liver.

As there is a contraction of the liver which renders it useless so there is another disease called **Cancer of the Liver.**

If the liver is overfilled with blood, individual obstructions easily form and as in all other blood obstructions hardening takes place.

These hardenings may increase and form the basis of Cancerous swellings; the more these develop the more incapable is the liver of performing its functions till at length death ensues; this is so because from the liver, by the Gall, the nourishment is operated on and the diseased matter is easily carried to other parts of the body where it again develops into Cancer.

This formation of cancerous swellings may last for a long time without the patient knowing or feeling anything of it; and when once this condition is perceptible to him it has generally gone too far for help to be of use. For the rest, the same applications and treatment may be used as in inflammation of the Liver.

Emaciation.

As there are people who have a particular tendency to Corpulency and who, even on the simplest food, look as though they were fattening, so there are people with whom the exact contrary obtains; they remain thin and lean let them eat and drink what they may. This leanness may be an inborn family trait so that all the members of the family, with but few exceptions, are thin.

Such people are as a rule healthy and strong and scarcely to be surpassed by the Corpulent. Equally are they talented and clever.

There is however a leanness which arises from disease or disposition to disease; these people look as if they were candidates for consumption or as if they might shrivel up.

As regards the thin people of the first class they have generally the best of appetites, they sleep well, and very rarely complain of illness.

It is quite different with those of the second class who complain of loss of appetite, a feeling of debility, of various bodily infirmities, of pains in the back, violent headache and pains in the feet.

To this class belong all those who, on account of various circumstances, do not get on. As a rule they cannot assimilate wholesome, strengthening food and therefore the whole machine is infirm and incapable of fulfilling its high duties.

Now, what is water able to do for such a patient? Those who experience no inconvenience from their leanness and feel strong and well with it I advise to brace their bodies so as to guard them against many diseases and to keep them in the best possible condition. I advise such to take two half baths and one or two upper washings in the week, beside walking from time to time bare foot in water, or taking a knee douche.

These applications will enable them the more easily to follow their profession and will give them a fresher and healthier appearance.

For thin people of the second class it is necessary both to work on the whole body, in order to brace, strengthen and confirm it, and also to act upon the special part so as to remove by degrees the chief mischief.

He who complains of gastric pains must see that his
stomach digests better and that by increased muscular
activity of the body all diseased matter is ejected: for
it is quite clear that either the stomach is out of order
or the inactivity of the system is so great that it is not
in a condition to assimilate and use up the diet properly. We often blame the stomach although it is often
quite innocent.

If once you get the system into greater activity so
that the perspiration is freer the appetite returns, a proof
that the suffering was caused by the inactivity of the
system generally and not by the stomach only.

A young man eighteen years of age complained of
constant stomach-ache, of oppression, of nausea and even
of actual vomiting. Withal he was so thin that one
perceived he could not be healthy.

During fourteen days he took one day a thigh
douche, another a back douche, and on the third a half
bath, and every other day an upper body washing extra:
the result was the vanishing of his gastric troubles, a
good appetite and a healthy appearance, a proof that
the system had not been sufficiently active.

In this example we see how great an advantage
the Water cure offers, and for the reason that it
operates universally and acts strengtheningly on the
organs so that special treatment is not always necessary.
Generally those people are very thin who suffer from
eruptions or from violent stinging and itching on the
whole body.

Some of them have appetites but others have none.
In these people there is certainly diseased matter which
contains a constant heat which burns or consumes too
much stuff, stuff which should have served for the
building up and maintenance of the system but which is
now diverted from its purpose.

There may exist a disposition to Consumption, or
great poverty of blood, which causes disturbances in the
heart's action or infirmities in parts of the abdomen

hindering the general prosperity of the system and causing emaciation.

In such systems a great deal can be done by means
of water, but one should begin with the gentlest applications in order to ascertain where strength is lacking
and where assistance is necessary.

A young man, twenty-five years old, was a frequent
sufferer from Colic. An extraordinary emaciation set in
which nothing would remove. He had sought the advice
of many Doctors and used many remedies without gaining more than slight temporary relief.

When he had gone through my treatment for four
weeks, which consisted of two half baths a week, a
compress consisting of a four-fold cloth soaked in hay-
flower water laid on the Abdomen and kept there for
an hour three times in the week and three teaspoonfuls
of tea of Centaury, camomile, and Milfoil every morning
and evening, the Colic completely vanished and a general
increase of bodily fulness appeared.

Thus Emaciation may be inborn, or developed by
various diseased conditions but whether it be from the
one cause or the other water properly used will render
valuable service in alleviating it or removing it.

A healthy nourishing food is often the chief necessity for emaciated people but it does not always produce
the desired result; still a good simple diet must be
persevered with, one that can be properly digested and
assimilated.

Corpulent people get back into a normal condition
principally by a dry diet and this is really best for the
emaciated.

Let us take an example from country folk.

As a rule these have the simplest food and are
fully occupied. If they adhere to their simplicity they
find themselves well off but if they begin the practise
of taking many superfluous articles of nourishment their
powers decay or they become too corpulent; indeed

many of those who were formerly like living skeletons
are now much stouter than they themselves approve.

The idea is quite wrong which formerly obtained
viz., **that a person was bound to take three measures of
fluid every day.** On the other hand that principle is
correct and confirmed a thousand times by experience
viz., "If you are thirsty drink, but do not increase your
thirst by the beverage. If you are not thirsty then your
food supplies you with enough fluid and you have a proof
that your blood is not watery but strong and sufficiently
nourished."

That habit plays the chief part here is not to be
doubted.

I have known people who never took fluid unless
they had been eating spiced or salted food. When I am
at home and taking the food I like best it frequently
happens that for one or two months I am not thirsty.
If, however, I take my food out even though it be only
a mid-day meal it is not more than half over before I
am very thirsty.

Malarial Fever.

There is a great difference between the air in a
deep valley and that on a sunny height, between the
air in a marshy district and that on high mountains.

Everywhere and at all times mists arise out of the
earth and according to the character of the soil so are
the vapours.

Naturally out of the marshes the vapours arising
from them are not so good as those arising from dry
plains.

As smoke is carried by the wind so these exhalations are borne in all directions by the air and we
breathe them in.

Just as dust is harmful to clothing and may indeed ruin it so it often happens with our systems; so also it happens that the various harmful exhalations which are taken by the breath into the human system often produce serious results.

It is quite clear that the evaporations from marshes take from the purity and healthiness of the air of those districts.

It is only necessary to visit the people who dwell on the lower coasts, in the boggy districts, in the stretches of land with marshy bottoms and exposed dried-up ponds and who constantly breathe in this evaporation so harmful to health, to see with what evils they are afflicted.

No one need wonder if in such localities fevers (specially Malarial) are the order of the day. It is impossible for those who live constantly in these districts to escape the fevers, of course those with sickly constitutions are the first to suffer. Malarial fever is a disease by which the whole system is reduced to weakness; it shows itself chiefly in loss of appetite, rheumatism in the limbs and exhaustion. Cold alternates with heat, in short the sufferer finds himself in a miserable condition. The attacks of fever appear at distinct intervals of time resembling in this respect intermittent fever. As regards its cure, one must operate on the whole body with the object of dissolving and extracting everything harmful that has insinuated itself into the system. And in this one will find help and benefit from whole washings which produce a general warmth and bring about the dissolving and extracting more rapidly.

Every day the whole washing should be made from bed every two hours in the manner recommended in "My Will".

In order however to attain greater effect and to get rid of everything bad in the system and bring it back to good healthy action it would be well to take, in addition to the whole washing, a half bath every second day and a knee and a thigh douche once during the week. For internal use I recommend three sorts of herbs viz. wormwood, juniper and pewter-grass all of which act on the stomach, liver and kidneys: the next best to these are Centaury-herb and water-trefoil, or angelica root with bark of oak.

The Measles.

Measles, like Scarlet Fever, belong to the frequently occurring childish diseases. The attacks of scarlet fever are not so frequent as those of measles: one rarely meets a child, especially in towns, who has not been ill at least once of measles. The inclination to this disease is sensibly greater than to scarlet fever.

Generally measles appear in children between the second and ninth year: they are more dangerous in the early than in the later years because as children get older their bodies are better developed and therefore more capable of resistance. Measles and Scarlet Fever are somewhat related to each other; in both the same symptoms appear, for example, fever, tears in the eyes and a dry cough.

By degrees spots as large as peas become visible on the face spreading after a day or two over the whole body.

During this time great care is necessary and a chill must be strictly guarded against.

In about three or four days the spots begin to blister and pass gradually into yellow places. I might almost call measles a mild scarlet fever for they differ only in individual symptoms. Measles are not so fatal but on the other hand there is the same danger of infection as in scarlet fever.

As in all these diseases fever exists so is it the case in Measles and the first step should be to subdue it and to effect this nothing is so good as ablutions in cold water.

If the children are still small they must be daily dipped rapidly two or three times in cold water and then immediately be put back into the warm bed.

Give bigger children a half bath daily and let them wear a shirt soaked in warm hay-flower decoction for an hour; immediately after removing the shirt give a whole washing.

By this shirt all poisonous matter hidden away in the system is extracted. I repeat here that the application should not · be made too often; every second day is quite often enough.

The best of hydropathy is that its operation is always towards the extraction of bad matter, strengthening the whole system and cleansing and regulating the circulation of the blood. It is wonderful how well chldiren recover after passing through an attack of measles if they have been carefully treated with water.

I must here warn you that it frequently happens that children who have been apparently cured by medicaments never properly recover; they are not exactly ill but they are not well and at length a fresh evil breaks out.

If the cure is set about in the right way the disease will be removed without leaving any bad results behind.

The regulations as to diet are the same as in Scarlet Fever.

~~~~~~~~~

Spleen.

On the left side under the chest lies the Spleen. It is not a real gland like the liver and kidneys but only a glandular formation.

The spleen, like every other organ, has a very high significance for the human body although it is a little difficult to grasp fully the importance of its task and to make it clear.

Very many splenetic sufferers have come to me for help and I have been able to cure them. With many of them the spleen was very swollen the result as I think of former severe illnesses brought on by blood disturbances. When these last were removed the spleen got into order therefore we may conclude that the spleen is of great importance in the formation of blood.

The enlargement of the spleen was quickly reduced by compresses on the abdomen made alternately with hay-flower decoction and water and vinegar. Compresses of oat-straw decoction are also good.

I recommend weakly patients to take four or five whole washings in the week and strong people two or three half baths and two whole washings, or instead of the half baths, thigh and back douche alternately.

The object of the water applications used in this disease must be to dissolve and extract.

As the spleen plays an important part in the formation of blood I know from experience how wonderfully herbal powders improve the blood and I recommend, specially for this disease, stinging-nettle, ribwort and sage whose property is principally to cleanse. They must be well dried, reduced to powder and a salt spoonful put into the soup which makes it taste like herbal soup: this should be taken for several days and the effect will be very good even as in other diseases. With spleen it is as with so many other sicknesses: the more sensibly a man has lived and the more his system is braced, the less power has the disease over him and the more the system is secured against the entrance of diseased matter from outside.

Atrophy of the Muscles.

He, whose body is in health and in order, with every part well developed, may be classed among the happy.

Such happiness does not fall to every one. Many suffer from a disease called Atrophy of the Muscles.

This disease may appear in various parts of the body, in the thumb for example, thus seriously affecting the movement of the afflicted member. The cause of this disease varies.

Very often symptoms of its approach appear after a severe illness especially when the recovery has been imperfect.

Atrophy of the Muscles may result from over fatigue. As far as I am concerned it is my firm belief that this disease is caused simply and entirely from poverty of the blood and insufficient nourishment.

It generally begins with fatigue of individual groups of muscles: more often it commences in the muscular region of the thumb; from thence it spreads to other parts of the body and the person experiences a good deal of pain. Neither medicines nor electricity will do any good here because they cannot restore the muscles. On the other hand I have, for many years, achieved great success with the water cure, and I never fear when such patients come to me because I feel sure of being able to help them.

The chief effort must be directed to distributing the blood in the proper way throughout the body, and the next to seeing that nourishment, capable of making good blood, is taken by the sufferer. Care must also be taken to secure good digestion and the best curative remedy will be found in tea of sage, juniper and wormwood.

The water applications should be quite mild so as not to excite the body.

As the decayed limb or muscle lacks power one must strive to improve the blood and juices and to

Kneipp, Codicil. 21

strengthen the whole organism. The atrophied parts
themselves should be wrapped every second day in a
cloth soaked in warm hay-flower water. If the patient's
constitution is strong douches should be begun at once
and in the following order. During the week a half
bath, a thigh, a back, and a full douche to be taken
and continued for about eight weeks when the applications may be reduced one half. Now and again the
atrophied limbs should be douched and the patient should
exercise himself moderately in the fresh air.

Nervousness, Neurosis and Weak Nerves.

There are three diseases which at first sight look
like three sisters, so much do they appear to resemble
each other, yet on closer observation they are found to
have special features and characteristics: these three
sisters are known as Nervousness, Neurosis and Weak
Nerves. All three are the result of weakness in the
human system which if removed would give no excuse
for their existence.

I can think of nothing more injurious to health,
strength, long life and endurance than the modern way
of living.

Now let us put the questions, where do nervous
people live? Where do these three sisters live?

In the Country, where one consumes simple food,
wears simple clothing, where people are braced and made
capable of resistance by work and good wholesome air?

No, there is no place for them here and they seldom
appear.

It is in towns or in those places where people live
after the fashion of cities that these sisters find a home.

If in passing through a town we could look into
the offices, Counting houses, scholastic Institutions and
trade workshops we should find them everywhere. Each

of the three sisters is possessed of so many attributes that it would be impossible to name them all.

It is difficult therefore to effect a cure because as the proverb says "Custom makes iron and the old habit of life will not be easily sacrificed."

Now whence come these three troublesome sisters?

This question is easy to answer: Mankind is getting further and further from the natural way of living and life is becoming in every respect different from what it formerly was.

As a rule country people consume a simple diet, require little beverage and in many families where life is still simpler no alcoholic beverages are taken at all.

With many people however the exact contrary is the case; the former go to rest at the right time for they appreciate the proverb "the morning hour has gold in its mouth," the latter on the contrary rob themselves of the best time to sleep and sit the whole night through at the public house with beer, wine or brandy.

The sensible man has his time for work and rest while the foolish man over-loads his mind with all sorts of business and allows his body to decay. We will make a comparison in order to see what becomes of these people. Compare a clerk or any other official employé with a simple farmer or countryman and we see the difference at once; observe a theologian, a priest and indeed all who occupy themselves with deep study; how worn they are before their studies are finished! Whence comes this? The answer is that their heads are continually occupied with thought and study, they take little or no exercise, neither do they take regular food: It is no wonder then if among this class of men there are many nervous and neurotic individuals.

They over-load their mind without considering the harm they are doing to their health. Naturally harmful results cannot fail; the whole human machine gradually gets into disorder and affords opportunity for one of the three sisters to take up her abode within it.

21*

Formerly I never knew that the business of a merchant was one that struck so deeply into the human system; it is quite clear to me now that there are many nervous people in this business. It is not only the activity and anxiety which are inseparable from the special occupation but the irregular way of living which brings the system into such a condition as to render possible the entrance of one or more of these sisters; still more the various entertainments and pleasures at the command of well-to-do merchants are all injurious to the human system.

Consider the factory workers; they must go to work early, and late take rest; their work is exhausting, the localities in which they work are often unhealthy; these, together with insufficient nourishment, form the chief factors in the condition of nervousness which is so widely spread among the working classes, a condition intensified by the fact that they generally choose a diet more exciting than strengthening.

These people often seek to strengthen their nerves by beer and spirits hoping thereby to work better for the use of them, forgetting that these drinks will prove their ruin. They have also the idea that bean-coffee makes the best breakfast forgetting that it is just the worst enemy of the human system. In my opinion nothing tends more to nervousness than bean-coffee.

If only I could induce these people to live more sensibly, to choose a stronger and at the same time a cheaper diet, and to avoid completely the drinking of beer and spirits they would be healthier and stronger and better able to accomplish their work in life. I must add here that if a person has had the misfortune to yield to his passions he must remedy it as soon as possible, for these passions never fail to bring sorrow to those who indulge in them and produce the greatest misery in their posterity as well.

The worst of it is that the people who indulge themselves in this way are, as a rule, very indifferent to religion so that the ruin is as it were double and their misfortune and discontent last for time and eternity.

If only those are really happy who fulfil the duties of their holy religion then are those doubly unhappy who have lost their religion, their faith, and their good habits.

What can a man do without religion?

Just nothing at all.

On the other hand how much is possible to one who is embued with the importance of religion and orders his life accordingly. With the one, happiness and blessing are at home, while the other feels forsaken and leads a discontented crippled existence I can imagine nothing more beautiful than family life when all the members are living near to God and leading a religious life for let the world offer what it may it can never make us as happy as religion can. If Masters of factories and employers of labour would live up to their belief and set a good example to their work people, letting their light shine, as it were, they would be astonished at the effect it would have upon them in rendering them happier, more content and quiet.

One point must not be forgotten. As a weakened, crippled body is more disposed to disease than a braced one, so over exertion of the mind leads not infrequently to mental aberration. Such a condition is a kind of cancerous injury to the soul. There are so many who find themselves in such a state that it is difficult to predict whether they will be overtaken by mental or physical paralysis.

The thoughts and ideas of many thousands are occupied solely with business and temporal things never remembering the most important business of all viz. care for the welfare of their immortal soul.

Country people who work hard are mostly disposed to religion; on the other hand wherever the three sisters have made for themselves a home, religion mostly gets pushed into the back ground.

I will not lay it down as a hard and fast rule but
I have found that where the three sisters live and reign,
there sensuality and passion are to be found and there
religion is set on one side: there is seeking and calling
after what the world offers, but internal rest and true
peace are not sought for.

Perhaps some of my readers especially those who
do not wish to know much of these things will ask
"What has religion got to do with nervousness and weak
nerves?"

How many thousands among the present generation
of men are crippled in mind and body! and whence
comes it?

Because they allow the most necessary thing to be
lacking viz Religion.

It often occurs that invalids suffer much more from
their minds than their bodies. There are many people
now-a-days who, as long as their strength holds out, not
only chain .themselves to their business but never think
of God or Religion.

They seek their joy and greatest happiness in temporal things and when their powers are crippled they
fall into depression, become neurotic and seek peace and
rest without finding it, for that which could really bring
these, Religion, they do not know.

Let no one tell me that Religion has no influence
on man especially those suffering from nervous disease.
For how many patients have been here who, after long
use of the water cure, have derived no benefit and whose
neurotic and neuralgic pains made themselves felt in
every possible place!

They could neither eat nor sleep, they fell into
melancholy and not till one knew the condition of their
soul and brought it into order did it fare well with the
physical trouble. At length they got rest and contentment and in short felt happier.

Let us take a look into a Country place where an unbelieving Doctor lives.

In such an one the people have no trust; if, however, they know that the Doctor is a good religious man they place double confidence in him.

Compare country children with school children in a town and look at them side by side and we need not ask which of these has made acquaintance with the three sisters.

Country children especially those who have an hour's walk to school in the morning and the same back in the evening and who perform this every day both in winter and summer are the strongest and healthiest.

On the other hand school children in Towns, weakened from head to foot, accompanied by a nurse, what do they look like? Very miserable and sickly and only in a small minority is there a glimpse of true health.

Go into a high school for Girls; in what does bracing consist here? My opinion is that the modern scholastic method is mostly designed to create nervous and neurotic people.

It is nonsense to look for these three sisters elsewhere than I have indicated; and those who have fallen under their influence through stupidity or thoughtlessness must at once look for a way of escape and a means of repairing the mischief already done.

There are three courses open to those who desire to be free viz., **bracing, simple diet**, and **proper care of the whole body**. Such people must commence with mild applications otherwise the exitement would increase rather than subside. They should begin by walking once daily on wet stones or in water. After a few days it may be done twice a day with an upper washing morning and evening. Again after the lapse of some days go back to the once a day walking on wet stones with the addition of an upper body washing and a knee douche.

So the applications go on slowly until one reaches full and half sitting bath, knee and back douche. The patient may divide these douches so that one a day may be taken.

All who persistently proceed with this treatment will gradually get into a better condition.

How very popular warm baths are in many circles! Yet from very large experience I am convinced that they are of no use whatever for these invalids, they only weaken while at the same time they excite the body. Many neurotic and neuralgic people have been to me after having taken from forty to fifty warm baths 28° to 30° but what had they gained by them? Simply nothing.

For example, I knew a military Official who had had over fifty warm baths; he was so weakened and excited by them that he would soon have had to retire. Before doing so he wished to try the treatment here in Wörishofen; he used the Water cure as prescribed by me zealously and in quite a short time he was completely well and resumed his duties with greater ease than before.

It is no unusual thing for invalids to use in a small way massage and electricity; knowing this, I have sometimes during my lectures asked my audience who among them had used these means of cure. In answer as many as fifty hands have been held up but they explained to me that they had been rather worse than better for the trial.

A Doctor who was lecturing to an audience of from eight hundred to a thousand began to explain and recommend massage as a means of cure at which there arose such a noise and opposition that he was obliged to leave the platform and was not a little ridiculed and mocked. Electrifying is very often recommended to nervous people and I will pass no opinion on it; but out of the large number of nervous people who have come to Wörishofen I have not seen that it has been of any service to those who had tried it, indeed it is incomprehensible to me how such an electric shock could strengthen the whole body.

By a neurotic condition I understand a general disturbance of the whole nervous system which shows itself by excitability, decreased capability, loss of energy, discouragement and melancholy.

People whose mental powers have been overtaxed are subject to this disease; it begins with pressure of the head, slight headache, quick beating of the heart, digestive disturbances, in short general bodily weakness.

I have frequently observed students who at first found not the slightest difficulty in pursuing their studies suddenly drift into so highly a nervous condition that they were quite unable to proceed with them.

A year before his "Final" a Student came to me and explained that he could no longer study and that he felt sure that the nervous headaches which he suffered combined with the anxiety he felt would result in his being "ploughed".

He used the water cure for eight weeks yet I could see no improvement in his condition. I advised him to combine field work with the water cure and to give up all study, and at the same time to feed very simply.

The Student agreed to this proposal saying it would be quite easy for him as his father was a farmer and would take him on as his third labourer.

At the end of three months he came back with the good news that his nervous troubles had quite gone. He then recovered his lost year of study with ease, passed his Examination and has been a priest for several years.

A man twenty eight years of age engaged in an office suddenly found himself unable to perform his work; he complained chiefly of a heavy head and insomnia but also felt thoroughly ill. He used the Water cure for

nine or ten weeks and although feeling better for it
was far from well. On my advice to try physical work
he engaged himself to a Gardener and tried gardening
for a small wage; while so engaged he lived on the
simplest food such as he had learnt to know in Wörishofen and used a Water application daily. Thus he lived
and worked for about a year; he found it difficult at
first but in time became accustomed to it.

I met him after six months and on enquiring how
all went with him he answered "Now I lack nothing."

Beside these I have mentioned many have followed
my advice and been the better for it.

Those who are nervous, neurotic, and melancholy
must not follow their own inclinations if the latter di-
verge from the path of simplicity and bracing. The three
sisters are like tramps who make acquaintance with
many people in town and country and become a burden
to them yet with perseverance and resolute will, they
may be completely got rid of.

In what manner these three bad sisters can be
shown to the door I have indicated already. If from
childhood one braces oneself properly, eats reasonable
food, and practises simplicity in all things, the three
sisters will never put in an appearance.

How many who once came to me in the most
miserable state meet me now overflowing with health,
strength and life and say "Since I have led the simple
life prescribed by you, I am quite well and no longer
suffer from nervousness of any kind."

Kidney Diseases.

The kidneys are of great importance to the human
body and have their seat right and left of the Vertebral
Column in the region of the loins.

The blood absorbs the matter prepared in the system in order to be increased and nourished, yet many things get into the blood which must be again ejected.

Again when the blood has made its circuit through the whole body and nourished all the parts it will contain much rubbish which must be removed.

The Kidneys serve therefore to disembarrass the blood of this or that foreign substance which has entered with the nourishment, therefore the blood is led into the Kidneys to be filtered and cleared of the same.

All that is useless to the blood they eject and let into the urinary bladder as water or urine. Marvellous as the arrangement is for the ejection of urine obstructions and disturbances will sometimes arise either in the Kidneys or in the Uretor and when this happens illnesses ensue.

Inflammation of the Kidneys.

Among the various diseases of the Kidneys that of inflammation is one of the best known and most difficult.

As inflammations may arise from chills in a general way so by a chill or like cause may inflammation of the kidneys be produced.

Then as usual the blood streams with greater power to the inflamed part and forms obstructions sometimes large sometimes small; the consequence is that the kidneys get overfull and swell.

The effect of this upon the invalid is that he feels first hot then cold, then suffers pain in the region of the kidneys; he has also difficulty in passing the proper amount of urine and suffers general disturbance in the body, all of which may end in dropsy.

Nor is this all, the appetite fails, thirst increases, because of the heat of the body, the powers begin to grow less, the breathing to get harder and the system becomes ever slower and more languid in its functions, and further progress would be marked by a general filling of the body with water and an increase of debility.

In such a condition there is a possibility of the Urine passing into the blood and corrupting it and so creating a very dangerous condition. In inflammation of the Kidneys one must endeavour with the greatest care to dissolve and extract the bad matter and combat the fever.

There is nothing better to use for a general effect on the body than whole washings, short bandages, upper and lower compresses and, if the invalid is strong enough to bear it, half baths also. In this case one or two of the former may be taken daily if the fever is well developed, and the half baths taken alternately with an upper body washing. By these applications the system is strengthened and able to perspire more freely and the bad matter is ejected and extracted. The more one can use the baths and washings the quicker will the inflammation be removed.

If the invalid is unable to endure the half baths, three or four whole washings should be undertaken daily according as the heat increases or decreases.

When the inflammation first sets in, the region of the kidneys should immediately be operated on. A plaster of pot cheese would be the most effective in this as in other inflammations, it should be stirred with pot water into a fine salve. The plaster should cover the whole width of the back and from the os Sacrum up to the shoulder blades. It is quite possible that one or two such plasters will suffice to remove the pain.

A further dissolving and extracting is obtained by placing the invalid three or four times a day on a cloth dipped in water and vinegar which should cover half his back.

This cloth should be re-dipped every half hour and not allowed to remain on longer than two hours.

A large double cloth which has been dipped in water and vinegar should also be laid on the Abdomen and renewed every half hour. Such an invalid could not bear the weight of a cloth of more than two folds, it would be too burdensome to him.

Remedies are used internally which act in drawing off the Urine and cleansing the Kidneys. The best remedies and most to be recommended are a decoction of rosemary which has been boiled in one third wine and two thirds water and a tea of brier, sage and dwarf elder root, a spoonful to be taken either of one or the other every hour. This tea mixed with a little wormwood or centaury herb would be of great benefit to the stomach.

Bright's Disease. (Kidney.)

It is not alone the liver that can contract: the kidneys are liable to a similar disease by which they become shrivelled and cause decay in the whole system.

When the kidneys no longer perform their functions properly bad matter stops behind in the blood and corrupts it, while on the other hand matter passes into the Urine which the system would find most useful.

In proportion as the Kidneys become diseased and useless, so the body grows infirm and injured; hard, cramped conditions set in until at last the mischief puts an end to life.

Unfortunately, strong healthy men attacked by this disease do not take notice of it until it is fully developed.

By degrees, however, their attention is compelled to their condition, for decrease of power sets in, an alteration takes place in the features, the healthy complexion becomes sickly, appetite fails, the sleep is irregular and the whole system is out of order.

As regards the cure and treatment of this disease the common rule holds good. If the mischief is not too far gone and the organs can still be braced and made useful then there is a chance of being able to effect a cure, if on the other hand the organs are too relaxed and the mischief far gone the whole system is ruined and cure impossible.

The applications most to be recommended in this disease are the thigh douche and the half bath for these act specially on the kidneys and benefit the whole body.

To lie for half an hour on a cloth or sheet many times folded and dipped in cold water is very effective in extracting diseased matter, nor must internal treatment be neglected if the body is to be cleansed and braced. The quicker the diseased matter is extracted and strengthening sets in, the more likely we are to get a good result.

Therefore I recommend, as specially suitable, wormwood tea with pewter-grass in change with tea of pine-bark and tea of juniper-berries and pewter-grass.

If this disease has made much progress the urine passes into the blood instead of through its proper canal. By this the brain becomes irritated and cramps occur. In this case a short bandage is advisable, if heat exists it should be taken cold otherwise it may be dipped in luke-warm water. Whole washings will also have a good effect.

Rheumatism.

Muscular-Rheumatism.

Certain seasons of the year develop special kinds of disease, for example, muscular rheumatism comes usually in Autumn and Winter and sometimes even in Spring. It generally announces itself when the change of temperature is very abrupt, when it is suddenly warm and then suddenly cold, or when the air is first damp and then immediately dry. In rapid change of weather one may get rheumatism in a few minutes and when once there it is very difficult to get rid of it.

For example, my Church was formerly quite damp as high as the windows, the walls at last were so decayed that they were quite covered with salt petre; the consequence was that when I read the service at a side altar close to the wall I regularly got rheumatism specially about the neck.

How did this happen?

The cold rushed or oozed out of the damp cold wall and struck me on the right side of my neck, suppressed the natural warmth and insinuated and established itself in the pores and little Canals. Thus the perspiration and extraction from the blood were pressed back and hindered and consequently an obstruction took place internally which made itself felt by stiffness and pain; in fact slight rheumatism had set in. I gave but little attention to it and after an hour's rest in my room I did not notice it.

Once I left a Window half opened in my room for about a quarter of an hour; the wind blew rather strongly through the open space and I soon found that my neck began to get stiff and caused me pain at every movement. This time it was very long before I was free of this very severe rheumatism.

It often happens that one goes to bed at night quite well and wakes in the morning with a stiff neck or a painful foot and one wonders how it could have come to pass; I will tell you.

During sleep the foot escaped from the bed and this short time was sufficient to set up rheumatism.

As the examples I have given show you the origin of mild rheumatism, so the more severe form originates in like manner; the cold air, penetrating by the pores, wanders further into the system and causes inflammation; this, in its turn, as we know attracts blood and produces obstructions and swellings which are extremely painful and difficult to remove.

As the swellings increase the pain increases and the stiffness gets worse. Muscular rheumatism can pass from one part of the body to another, from the arm to the foot, from the foot to the body and acts prejudicially on the internal organs; if it draws near the heart it is dangerous.

The invalid attacked by rheumatism does not always have fever but none the less the pains are often very great while for the rest he feels healthy and strong.

This disease often lasts weeks, months, and even years, then hardenings or swellings form which are difficult to remove.

If only single parts are attacked and the rheumatism still wanders the disease is not so important.

If however, it spreads to the greater part of the body and several hard swellings have formed, then not only does the pain increase but the stiffness of the muscles also, so that the sufferer can no longer move and shrieks with pain if one only touches him slightly.

The pains are not continuous yet all complain of violent twinges which pass like electric shocks through the muscles attacked by the disease as if the telegraph wire led from one place to another.

The stiffness sometimes increases so much that the patient is only able to walk with a stick or on crutches and the whole body gets into a troubled state.

Not only is rheumatism brought on by sudden changes of temperature; quite as frequently it is the result of draughts, damp houses, new buildings or damp climate, and especially is it caused by debility which makes the entrance of all disease into the system so very easy.

Certain professions in life play an important part in this disease because so many people in the exercise of their trade or profession undergo a rapid change of temperature, for example the brewer has to be busy now in cold, now in heat, now in the cellar, now in the brewery.

It is almost impossible to enumerate all the causes which produce and favour this disease.

The cure of muscular rheumatism is extremely difficult and medicines are powerless here; for if mild drugs are taken they do no good and if strong ones they tear down more than they build up.

Here is my opinion; Whatever has penetrated the system and done it harm must be extracted and the simpler this is performed the quicker will be the cure.

For example a person gets a very stiff neck by a draught and he can neither turn it to the right nor to the left; it is as though he had a hard and painful block in his neck; if he will lie in bed, wash his neck every half hour with cold water and, without drying it, bind it about with a dry towel he will find benefit because the warmth caused by the washing will be absorbed by the cloth and begin the extraction.

He should repeat the washing and the dry towel every half hour and see that the latter covers all up without being tight.

This increases the warmth and produces a vapour or steam which is nothing else than the extraction of the bad matter; even the head at times perspires and should not be discouraged.

After three or four hours of this treatment the neck becomes flexible and soft and in a very short time will be free of rheumatism in consequence of the harmful matter being either dissolved or extracted.

As in the neck, so also in other parts of the body rheumatism if it be not deep seated may be removed by washings and dry bandages.

If however, the rheumatism is of·long standing then the above applications will no longer suffice and a stronger method is necessary such as bandages of warm pewter-grass water or hay-flower decoction taken in change with whole washings and douches of cold water.

Supposing the rheumatism to have spread to several parts of the body these parts must of course be operated upon but at the same time applications must be given to the whole body.

A person I knew had a very stiff thigh and was full of rheumatism throughout. It was supposed in this case to be sufficient if the operation was confined to the affected bone in the thigh and not to the whole body; but such was not my opinion; I considered it needful to remove the disease by general treatment, by thigh douche with half bath and by full douche and full bath, and ordered the following applications to be made in the week: three thigh douches, two half baths, one back douche, and a full bath.

For even admitted that one could cure the thigh alone, still the remaining parts of the body had undoubtedly suffered by the disease and lost some of their former power and activity and were in no way protected from a recurrence of the evil.

I know well that there are hot mineral springs and that these and hot baths are prescribed for the cure of rheumatism. I have even tried such experiments myself and have come to the conclusion that warm baths are powerful dissolvers and extractors but they so weaken the body and impoverish the juices and the blood that the rheumatism returns on the slightest provocation and trifling change of atmosphere.

At most it would not do to take more than one or two warm hay-flower baths alternately with a cold full bath for I have tried it.

Experience has shown me that such warm baths are not at all needful because by the bandages and cold applications rheumatism can be completely and easily removed with all the harmful matter which has taken up its abode in the system.

As the douches and baths must be only of short duration so also the bandages must not be kept on long, for if the bandage is kept on after it gets hot the rheumatism increases and the inflammation grows stronger so that at last one has done more harm than good.

Because the application of bandages requires such great care I have, taught by experience, confined myself largely to cold applications, such as douches, for the treatment of rheumatism, and have never regretted it.

The internal operations for rheumatic conditions are very simple viz., good nourishing diet which will form good blood and in order to secure good digestion see that the bowels act regularly.

If good appetite and proper action of the bowels exist it is not needful to take other measures; should the former be lacking it must be assisted by a suitable tea. If the latter be out of order take a spoonful of water every half hour and, if this be not sufficient, sit once or twice daily for ten minutes on a cloth, several times folded, which has been dipped in quite cold water.

A very good tea is made of sage with water-trefoil and wormwood or of centaury-herb, ribwort and Milfoil or of Angelica root and centaury herb or finally of furze, heath and wormwood.

Rheumatism in the Joints.

Rheumatism in the joints is distinguished from muscular rheumatism in that it has its seat in the joints where all the harmful matter brought together by rheumatism collects and settles.

By thickening and hardening a compact mass is formed which completely prevents the activity and use of the joint which is, if I may use the expression, walled in.

Rheumatism in the joint may be two fold that is acute or chronic.

The former develops as quickly as any other inflammation and may last for weeks and months if not

22*

combatted vigorously in order to dissolve and extract
the evil matter.

A person attacked by this form of rheumatism feels
in the joint, especially in the right knee, a burning pain
which increases till it becomes intolerable, fever ensues,
the joint begins to swell greatly, the skin gets red and
shining and soon he cannot move the joint.

If the disease is making rapid progress it goes from
one joint to another till he finds it impossible to move
and suffers fearful pain. One can scarcely imagine any
condition more pitiable: it is plain to see from the man's
face how much he suffers and the attempt to move him
causes him involuntarily to scream. He seeks a position
in which better to bear the pain but cannot find it; his
appetite fails and his thirst is so great that he knows
not how to quench it, his sleep is broken and the action
of the bowels very difficult and he has great touble in
passing Urine which is corrupt and mixed with much
impurity, and there is every indication that the digestion
is out of order and that the circulation of the blood is
imperfect. This condition may last for several days or
even weeks; the mischief encroaches on the internal
organs and frequently acts prejudicially on the heart.

As regards the cure one must clearly distinguish
whether the invalid is very emaciated or possessed of
a certain amount of strength; in the latter case the
application may be stronger. If, however, natural warmth
be lacking and the general condition very weak one must
commence with the mildest treatment possible.

Such an invalid can only be brought to an equable
warmth and regular perspiration by whole washing made
once or twice a day; aud the places severely attacked
by rheumatism must be specially douched once or twice
a day.

If, on the other hand, the patient is in possession
of a tolerable amount of natural strength and normal
warmth he should take a whole washing or strong douche
once or twice a day and treat the specially painful limb

or joint with douches but this will be seen best by the following examples.

A youth eighteen years old had for a long time suffered from rheumatism in the joints of both knees and came here to Wörishofen to get help. We began by giving him a washing daily beside a knee douche and thigh douche on alternate days these rendered the system more active, sometimes we gave him a back douche instead of the thigh douche. We continued these applications for about five days then we gave one day a thigh douche, the second day a half bath, the third a back douche and the fourth a full douche. The effect of the daily douchings was to decrease the pain and the inflammation and to cause the whole system to lose its superfluous heat.

If however the **joint rheumatism** is of extraordinary strength and the system tolerably strong at the same time then this is the simplest and surest treatment: douche the affected parts as often as the pain comes on even though it be seven or eight times a day; let the douche last until the pain ceases.

As in this way individual parts of the body are handled so the whole body must be treated if the rheumatism of the joints has spread over the whole body.

A railway official who suffered from joint rheumatism had douches daily beside other smaller applications; at first his condition slightly improved then the improvement ceased and the rheumatism was worse than at first.

When this had gone on for six weeks I was called to his sick bed. The man suffered such fearful pains in his feet, thighs, shoulders, and in his whole body, that he cried aloud and had already given up hope.

I ordered the man who was well built and otherwise robust to be carried by two men to the neighbouring bathing tub in spite of all his pained and woeful cries and douched him with ten garden hoseful of the coldest water beginning at the feet and then gradually mounting

upwards.՚ Finally at the end of the tenth douche he
said "I am free of pain".

At once I stopped douching and the patient was
put to bed again and was unspeakably happy because
his pain was removed. I gave the advice to repeat the
douches in the same way as soon as the pains returned
even if they should have to be repeated seven or eight
times a day. I was obeyed and on the tenth day after
I had given him that powerful douche he appeared before me at the "consulting hour" dressed in his uniform
and free of pain and no one believed that this was the
man who had been lying so ill a short time before.

Had he gone on with the one or two douches daily
he might not have god rid of the rheumatism for weeks
or even months because the heat would have developed
between one application and another establishing the
rheumatism rather than getting rid of it.

But as the douches were repeated as often as the
heat and pain returned the power of the disease was
broken and as we saw, the man lost the pain completely
even on the first day.

It is very important therefore to meet every setting
in of the pain with a douche.

The internal treatment is similar to that recommended for muscular rheumatism.

Chronic Joint Rheumatism

is not a passing condition like the **acute** which comes
perhaps once or twice a year and again leaves; but it
is continuous because it establishes itself in the joint
and forms obdurate swellings which are rarely if ever
dissolved or removed.

Invalids who suffer in this way have a very pained
look, a pale complexion, sunken features, they cannot
sleep, have little or no appetite, their body has lost all
healthy activity and the whole system is sick and ill.
The pains are as a rule very severe although little or
no fever exists.

The changes of weather have a marked influence on these sufferers; they know beforehand whether the weather will be wet or windy, for at every change they have violent pains, or burning and stinging in the joints attacked by this disease.

What then is to be done?

The hard swellings are a sign that the used up ingredients in the system have not been extracted. Therefore in the treatment one must secure the dissolving and ejection of the diseased matter which has helped to form these hard swellings.

When this is done the normal condition sets in again in the individual joints as well as in the whole body.

If only single members are attacked by the rheumatism I consider it best and more to the purpose to treat it with bandages of hay-flower, oat straw, or pine-bark decoction.

They may be kept on from three quarters of an hour to two hours, if, however, they exceed three quarters of an hour the cloth must be redipped in the decoction otherwise it would develop too much warmth and increase the mischief, or the action of the bandage might cease and then cold would set in which would be equally harmful.

Too many bandages must be avoided.

However needful warm bandages are, the cold applications must not be neglected, for in this disease the whole body suffers and must be strengthened; and as the cold water heals and dissolves, the cold and warm applications should be taken alternately.

The bandages must be arranged according to the strength or weakness of the patient; if tolerably strong he may have a bandage once or twice a day, one or two douches on the suffering limb and twice a day an application on the whole body. In great weakness a washing, or thigh, or back douche or half bath will be sufficient.

Consumption of the Spinal Marrow (tabes dorsalis).

Country people rarely suffer from this illness but it is not infrequent in the higher ranks.

It is often a consequence of other diseases especially of Syphilis the poisonous matter of which penetrates into some part of the Spinal marrow and spreads further. It happens in this special place where it has penetrated as if a little ulcer arose and would not heal and spread itself, disturbing the organs.

The leaves fall off and the branch withers.

The effect of the mischief which began in the Spinal Marrow is felt throughout the whole body, the damage spreads, the body lacks nourishment and decay increases.

The following are the signs of the presence of this disease.

Strength decreases, uncertainty in the movements, especially of the legs, sets in, the feet refuse their service so that the invalid scarcely dare step, sensation loses its acuteness, an upright position becomes impossible, palpitations set in, the rest is so disturbed that sleep is out of the question, in short the whole organism is out of order.

The pain is neither burning nor stinging but shoots like lightning through the limbs especially through the legs. In this way the misery increases, the strength diminishes, emaciation takes place, the Urine passes with difficulty or not at all, the bowels do not act, in a word there is no longer harmony in the working of the bodily organs and at length death sets the sufferer free.

If the disease has only just begun and the system is still robust the bad matter can be removed by powerfully dissolving and extracting.

The quicker the interchange of stuff goes on, that is to say the quicker the used-up and diseased stuff is ejected and the better nourishment administered, so much the surer will the evil be removed.

First of all then the system must be brought into proper working order, the natural warmth must be increased and the whole organism strengthened; then very soon a good appetite, a free perspiration and a general strengthening will establish themselves.

Upper douches will quickly bring the body into greater activity, the powers will increase and everything harmful will be drawn out.

Thigh douches are equally strengthening and bracing, the flaccidity will vanish and thus the diseased matter will be fought effectively on two sides.

If the upper and thigh douches have operated well then the half bath will strengthen the whole abdomen still more, it will produce greater activity and increase the normal warmth. The back douche extracts the diseased matter and thus by degree the system gets into order.

Further try for a good digestion so that the diseased stuff may be properly ejected. Wormwood tea with centaury herb and sage strengthen the stomach and improve the blood.

Tormentilla and Angelica root tea dissolve and eject poisonous matter; and in order to build up and strengthen the weak parts of the body and to act upon the bones I recommend bone powder as the best remedy. Bark of Oak tea acts bracingly on the whole system; it must be taken in moderation otherwise it will cause constipation.

Of course such a disease as this cannot be cured in a few days, especially if, as is often the case, the germ has existed for ten or twelve years and already developed into mischief.

Combined with this disease there is often severe constipation as well as disturbance in the ejection of Urine.

The action of the bowels is best regulated by a spoonful of water taken scrupulously every hour.

If preferred a tea spoonful of weak wormwood tea may be taken alternately with the water.

If this is not sufficient it would be well to sit once or twice daily for ten minutes on a cloth soaked in cold water. The lower compress also has a good effect in this disease because it braces the system, decreases the heat, and causes activity in the Abdomen. It should be applied two or three times a week until the action of the bowels becomes regular when it should be discontinued.

For extracting or drawing off the Urine a compress of pewter-grass-tea on the Urinary bladder and a rather strong tea of oat-straw, two or three spoonsful daily, are very beneficial.

The special pain of this disease is not as a rule continuous nor confined to one place but goes like lightning through the body from one place to another, above all down through the legs; this being so the legs should be rubbed two or three times a day with a decoction formed of half water and half vinegar in which Arnica has been boiled.

There came to me a man in his fortieth year; he staggered like a drunken man and could with difficulty keep himself upright as he came up the road, at length he collapsed. He was almost powerless and very emaciated, he had scarcely any normal warmth, was quite sleepless and greatly feared water.

Several specialists and well known Doctors were agreed that this patient had Spinal Marrow Consumption in a high degree.

His Wife accompanied him and at once put the question "Have you studied medicine?" My answer was "**No**".

"Can you attempt" asked the wife further "to cure my husband who is all I have in the world? We have travelled for ten years from one celebrity to another but have found no help."

The Doctor declared him to be incurable and required that I should dismiss him.

From pity and because I had still a little hope I made the proposal that a five days trial should be made of the gentlest applications.

I explained to the wife that she and her husband had come to me of their own accord, I had not sent for them, that what I intended to do would not hurt the man, still if they wished to go I had no objection.

The man and his wife remained.

I first douched him from feet to knees with the coldest water and indeed I rubbed his feet with my wet hand until he was thoroughly wet. At the second application which took place on the same day I douched his body up to the region of the stomach in the same way.

On the second day I first douched his feet as before and squirted half a garden hose of water over his knees and somewhat later in the day the upper body and back were washed.

On the third day, I gave him a whole garden hose full of water over his knees and douched his upper body and back.

At this point he got warm during the night and slept better than he had done for ten years.

Thus the cure was continued, from knee douche he went to thigh douche and then to upper douche. The warmth increased tolerably fast and with the warmth came strength.

In the same way appetite and sleep increased. After ten days came upper douche with back douche, upper douche with thigh douche and walking in water alternately.

To the man who had formerly believed cure impossible came **hope.**

He acquired daily a greater love of water and the
applications became necessities to him. This treatment
was continued for six weeks at the end of which time
his appetite was good and his sleep refreshing.

He left Wörishofen and came again after a year but
he walked so well that it was difficult to see that he
still had a trace of the disease.

He again submitted to treatment went away and
came yet a third time but only to walk up and down
the Village and show himself, for he now enjoyed the
best of health. His wife, however, whose health had
been formerly good, had now departed this life.

Delirium Tremens.

There are people so addicted to Alcohol as completely to ruin body and soul; they have gone so far
that they cannot exist, neither can they work without
their beer or brandy.

Immoderate drinking is a fearful burden for the
body to carry about under which it sooner or later col-
lapses. These people begin at first with small quantities
which increase day by day and intoxication which was
formerly of rare occurrence becomes a daily condition
and eventually paralyses every power both of mind and
body. Reason and sense are disturbed and free will is
so hampered that imbecility and in some cases even insanity results.

How can these people be helped?

The first thing must be to avoid this vice. And
as such men are not masters of their will it is necessary
that they should have a good attendant about them.

Intemperance does not allow itself to be removed
suddenly indeed it might do harm even if it were possible; therefore one must proceed a step at the time to
get rid of this frightful habit; to begin with, take

simple strengthening food, practise bracing and make suitable water applications. There are none better than knee, thigh, and upper douches with full and half bath alternately, combining each day a strong with a mild application.

In this way the unfortunate person may get back into a normal condition.

Internally, tea of ribwort, milfoil and wormwood would be very good, so would tea of Angelica root tormentilla and ribwort.

Examples are known to me that the intense yearning for alcohol which makes itself felt in these people not infrequently vanishes within a month if this treatment is followed.

Thus two men came to me one day; the father presented to me his son who, by his immoderate drinking, had completely ruined himself.

The Doctor told him that although he was only twenty-four years of age he would die if he did not give up drinking.

The son brought the same charge against the father and both deplored their trouble to me and showed into what a pitiable state they had fallen.

I said to them "If you completely give up this vice and use a suitable water cure then help may partly be given to you.

Both promised to do so and really fulfilled this promise; they chose only the simplest food, avoided all alcoholic drinks, used two water applications daily and after eight weeks they were looking so well that one could only wonder at the marvellous effect of water.

Scurvy.

There are very few people who possess pure clean blood: this is perceptible more or less according to the individual. Those who have the purest blood should consider themselves happy while those whose blood is corrupt may be rightly considered unhappy people. For who can enumerate the many diseases developed, in the system by bad blood!

Scarlet fever, measles, eruptions, herpes are all the results of bad blood, but one more than all others is the result of impure blood and that is scurvy.

Scurvy is a disease which overcomes man by degrees: at first he can quite well follow his calling but as it gets power he is unable to rise from bed.

It makes itself more noticeable internally than externally; internally by pains, externally by brown or yellow spots. The gums swell, the teeth get loose and the breath is evil smelling. These are all signs of great infirmity in the whole body. Not only in the gums do ulcers form from which a foul pus issues but in the body also. Naturally a man suffering from Scurvy is very feverish. The question is what can one do in such a disease seeing that the whole blood is corrupt?

Certainly Medicines will not suffice for a general improvement of the blood and extraction of bad matter. As the whole body is corrupt it is quite in accordance with nature to operate on the whole body in order to get rid of all that is unhealthy. Attention must also be paid to the ulcered mouth.

Cold water applications are best and during the week a shirt dipped in hay-flower decoction, a full bath and a half bath are necessary. The former application both dissolves and extracts.

As further applications there might be an upper body washing with cold water and a knee douche beside, during the day. If these are continued conscientiously for twelve or fourteen days the disease will disappear.

The best internal remedy is tea of centaury herb, fennel or sage.

To get rid of the ulcers in the mouth it is necessary to gargle throat and mouth frequently with a decoction of pewter-grass tea.

Although this disease is difficult yet it may be got rid of if one knows how to operate in the right way. It must be borne in mind that even when the system appears to be reinstated the cure is not entirely accomplished; to really effect this the body must be braced for some time longer.

It is only by degrees that full health will return. The directions I have given as to treatment are general; it is the business of each individual sufferer to test his condition as to whether the applications should be taken weaker or stronger.

Scarlet Fever.

Scarlet fever is a very dangerous illness to which many children fall victims.

The disease is most prevalent in winter or between Autumn and Winter and spreads with such rapidity that in a few days a whole village or district may become infected.

Panic attacks both big and little people; schools are closed and mutual intercourse is avoided as much as possible.

It is a disease more general among children though unhappily adults are not free of it. And it is one that leaves so much mischief behind that many lives have been ruined by it; for example, some have lost their hearing, others their sight, while in others it has left dropsy, bone ulcers and so on. It is a disease which leaves immense mischief behind specially if it has not been treated in the right way.

I am fully convinced that by treating scarlet fever in the proper way, that is by extracting the bad matter by means of cold water, the disease loses much of its malignity.

Between the age of two and eight children are specially subject to it.

The cause cannot always be traced, it may be by breathing in the corrupt air of the sick room, or it may be conveyed by means of articles of clothing or bedding used by scarlet fever patients and often one cannot tell how the disease has found entrance.

The first symptoms of it are languor in the whole system, violent fever, rapid pulse and heavy head. In this condition the invalid remains for some days when a further symptom appears in an eruption of red spots on the body.

With this the danger to life reaches its highest point: the patient is in a continual fever and devoured by thirst. The red spots last for seven or eight days and then begin to fade and disappear.

Now I ask, What is to be done with such a patient? How medicines can extinguish such a conflagration as is to be found within him is inconceivable to me!

Here cold water is the sole and only remedy for stopping the fever heat and cleansing and bracing the system.

The moment the symptoms appear be prepared with your attack. If the child is in high fever with heavy head and loss of appetite put on without delay a shirt dipped in cold salt water, wrap over it a blanket and cover well up though not too heavily. Let the child lie in this shirt for three quarters of an hour, then remove it and put him or her back to bed.

By this shirt the bad matter will be brought to the upper surface of the system and the child will experience decided relief. After this the child must be quickly washed every two hours with cold water and placed back again in bed. Now and then it would be well to give a cold half bath lasting two seconds only.

Continue this treatment until the fever has gone and the evil condition is removed. If the fever shows any sign of returning let a whole washing in cold water be given.

Instead of dipping the shirt in salt water it may be dipped in simple cold water and laid on the patient who should be covered with a blanket and again put to bed.

After three or four days a marked improvement will appear both in appetite and in physical condition.

It is remarkable how the patient prospers who is treated in this way; the reason is clear; it is that water ejects bad matter and cleanses and braces the system in a way that medicines could never do.

Beside this external treatment give the invalid every hour a spoonful of water to allay the thirst and small portions of tea made of sage and ribwort or of fenugreek.

Whether the patients are children or adults the treatment is the same. If, while suffering from scarlet fever, the patient is attacked by Ague put on a shirt dipped in warm hay-flower decoction instead of one dipped in cold water. Avoid panic or fright if you hear of an epidemic approaching.

Let those attacked by scarlet fever be at once isolated and see that the sick room is well aired and the linen and clothes disinfected.

To allay thirst give water on the teas I have named in small quantities but very frequently.

The food must of course be simple such as I always advise in sickness.

If my advice is scrupulously followed one need have no fear for patient or attendant for even if the disease has spread it will soon disappear.

Perspiration.

Breathing is a continuous business for every one in the world by day and by night, in sleeping and in waking.

By inhaling or inspiration the system gets nourishment from the air necessary to its life and maintenance.

Just so by expiration or breathing out the used up matter in the system finds its exit. The inspiration and expiration which goes on in the whole body is so gentle that we scarcely notice it. If the breathing stops or is hindered in some forcible way life is no longer possible.

Any interference with the breathing causes disturbances in the system and, by preventing free perspiration, produces disease.

If a cloth dipped in water is hung in the air it steams all the watery matter out of it and the air absorbs it.

Just so it happens in the human body; if fluid issue from the perspiration glands and steams into the air that is nothing extraordinary. If, however, severe trials of strength or increased activity exert the body more than usual a larger quantity of sweat or perspiration issues from the pores, so that it not unrarely runs down over the face and body in great drops. This sweat is pressed out in a forcible way, because the extraordinary activity and trial of strength have caused an increased warmth of the body.

A similar effect can also be produced by anxiety, fear, fright, joy and sorrow on the ejection of sweat: one has often noticed that a sudden fright has caused drops of sweat to appear on a man's face and body.

The effect of fright and anxiety upon children is that they cannot retain the Urine and that the action of the bowels is involuntary.

There is however another sweat or profuse perspiration which does not arise from exertion, fear or anxiety.

I might describe it as caused by a duel between life and death in the system. A person is very ill, the question is which will conquer, the constitution or the disease?

If the patient perspires it will be the Constitution which is victorious and the disease which is vanquished. This is generally called the critical sweat, that is, critical to the welfare of the system because as a rule it is the first sign of recovery.

I know of yet another sweat which is undoubtedly the worst and in which the system not only lacks heat but all its functions are crippled and its activity fails, then it is that a fluid stuff issues from the pores as **cold sweat** which is detrimental to the patient and is a precursor of death.

Perhaps my dear Readers would like to know how sweat or perspiration arises and how it is regulated! To make this clear let me introduce an example.

I know a rather large eminence or high part of land in which rise several springs and if one makes a hole or small pit with a pickaxe or hoe, up bubbles a new spring.

Inside the hill are canals through which the water has made exits for itself. Exactly so there are in the human body similar canals called by us perspiration glands and by which the fluid makes its exit. If, by hard work or nervous influence, any pressure is exercised then the sweat is forced out and the drops flow in streams over the face and body.

If then many such influences are brought to bear on the body the perspiration glands empty themselves; the system then gets poorer in fluid and after profuse perspiration there is great thirst simply because the lost fluid must be replaced.

As the Canals can always be filled so can they fall in and dry if a lack of fluid exists or if the system is not properly active. Should this be so then all is wrong with the man.

23*

It is most necessary that the Canals should be kept in perfect working order by free healthy sweat or perspiration which takes in to the system nourishing matter and carries off all that is used up and hurtful.

One often hears the complaint "I have not been able to perspire for a long time and I am in consequence uncomfortable" or „If I sweat I sweat away the disease".

From what has been said it is clearly necessary that the used up matter should be ejected by sweat.

As a rule the perspiration glands empty themselves without difficulty, yet in individual parts of the body sweat may appear in large quantities while other parts remain quite inactive.

Several patients have come to me with the complaint that half their bodies perspired freely while the other half was always dry. Again I have seen some with drops of sweat on one side of the face and not a drop on the other.

Just as half the body may perspire and the other half not, so individual parts may perspire more than others. There are, for instance, many people who sweat the whole day underneath the arms; whether they work much or little it is always damp there. Damp or perspiring feet mean nothing else than a strong ejection going on there.

Here the sweat may be so strong and biting that the skin is eaten away, ulcers form and a sort of corruption sets in which emits a sickly smell.

There are again many whose hands are constantly damp and from whose fingers drops of sweat fall; others perspire only in their heads and should this cease it causes them discomfort because obstructions form which hinder ejection. Beside the perspirations already indicated they occur in places not always noticed but still having results.

The more faultless a machine is the more perfect is its work; if disturbances occur in it it goes badly. It is precisely so in the human system.

The better and more completely the system is provided with necessaries and the more active the Canals are to eject and extract everything deleterious so much the better and stronger is the man; on the contrary if the machine is not in order and injuries not set right bad consequences must result.

In Consumption and some other diseases severe sweat occurs on the slightest exertion but it is equally prejudicial if the necessary ejection of matter does not take place.

The best proof of this is that if such a sweat is interrupted or suddenly stopped disturbances take place directly; therefore great care is necessary that the sweat is not forcibly removed; the thing to do is to try and operate on the system so that it shall be in a condition to receive the good and eject the evil.

To bring about that the sweatings should be in accordance with the constitution it is above all necessary that the whole system as well as individual parts should be brought into the best condition for I need not say that every part is sick which does not properly perspire.

Thus the most important thing is to strengthen and brace the whole so that all the parts are equally active. Therefore the canals which are weak and dry must be awakened, revived and brought into action, the more effectually this is done the better will it be for the whole system.

The applications which bring about and maintain the best order are the upper baths with upper douches for they give strength, bracing, and activity to the system.

For delicate people whole washing alternately with the bath or upper-body-washing with thigh douche I strongly recommend; in a word, such applications should be used which are capable of retaining the normal warmth, vivifying the whole body and bringing into action and strength every organ.

The weakness which very often follows an illness and may prove dangerous is best removed by gentle applications of cold water and strengthening teas. To strengthen the system is the one great thing to do here.

Among the various teas I have found sage with centaury herb or sage alone to be best. If one takes a cup daily in two or three portions, or a spoonful every two hours it is sufficient.

Bark of Oak with sage may also be recommended; wormwood, centaury herb and sage act bracingly on the stomach and improve the juices. If, during illness, the invalid gets into a strong unbroken sweat he should take a whole washing in cold water; if he is strong enough let him have a half bath; if he feels weak then a whole washing from his warm bed will suffice.

In any case whether it be the half bath or whole washing it must be accomplished as rapidly as possible; the former should not exceed two seconds and the latter should be over in one minute; if they occupied a longer time the body would get chilled and rheumatism would probably set in.

Sweating feet must be specially treated.

One must not think that all that escapes from them is from the feet only, it is more than probable that the corrupt, foul fluid has been forced down through the Canals; therefore the sweat from the feet is matter from the body ejected through the feet. If one tried forcibly to suppress this sweat and close this opening, he would rouse and drive back the bad stuff to the body and so cause obstructions which might end disastrously; they lead to dropsy or other diseases dangerous to life.

This sweat can, however, be cured by operating on the whole body because equable perspiration would result and all used up stuff be properly extracted.

If any one is troubled with sweating feet let him during the week take two whole washings, one short bandage and one half bath, then such activity will set in in the body that sweating feet need no longer be feared.

The feet should three times in the week be wrapped in a cloth which has been dipped in a warm decoction of hay-flower or pine - bark and remain so for three seconds each time; thus the dead skin and corrupt pores are cleansed, healed and braced.

This is really all that is necessary in order to rid oneself of sweating feet.

I advise very strongly that those people who have suffered from such a complaint should, in summer time, walk bare-foot on the grass and in water for one or two minutes at the time; and even in the Winter in fresh fallen snow for by these bare-foot exercises the system is thoroughly cleansed. As I have said before the first and all important step is to get the body into a healthier state and enable it to eject all used up stuff and prevent the entrance of foul matter.

Sea-Sickness.

Travellers who come to me from over the sea have all something to say about sea-sickness.

It is a disagreable condition that few are spared who go to sea for the first time and while suffering from it they feel so very miserable that they would almost welcome death.

This illness is not exactly dangerous yet it may become so if any other mischief is brought on by it.

Sea Sickness depends on changes of temperature and the evaporations which arise from the sea but specially upon the rolling swinging motion. I advise all who cross the Sea to take a good meal before going on board for an empty stomach yields to the attack much more readily than a satisfied one. I have given to many of my patients an antidote to sea sickness which they say has been of great service. It consists of a tincture of herbs which has the effect of supporting the stomach and preventing or removing nausea.

A couple of washings and one or two half baths during a week's voyage would also be of great help.

I have often been asked whether sea water is as good for bathing in as fresh water. In reply to this question I may say that many have come to me after taking sea baths because they derived no benefit from them: But is this the fault of sea water? No.

The fault is that the sea baths are much too long. In consequence of the long duration the salt water exercises a more or less prejudicial effect on the system whereas this is never the case with a short bath of at most two minutes' duration. I have also been asked if hot sea water is good for bathing. I can only reply that water always forms a good bracing remedy and is never without effect. Yet I give preference to the coldest water possible, for the colder it is the better it acts; on the other hand the warmer it is the more it debilitates the system.

Tetanus. (Lock Jaw).

In my home a strong healthy lad was shot by a huntsman in the hand; unluckily he paid no attention to the wound which ought to have received great care and in a short time **wound cramp** came on which, in the vernacular, is called Canine spasm. The poor lad suffered horribly; the pains having the character of convulsive cramp.

The pains spread rapidly over the whole body and by degrees a complete stiffness set in.

In Tetanus it happens in a certain sense similarly to blood poisoning. In this generally a wound has existed, through which bad matter has pressed into the blood and caused inflammation. This continues to spread till all the blood is corrupt and death ensues. In Tetanus also a wound generally starts it and by preference in the hand.

At first one scarcely notices it until a convulsive pain sets in which is called forth by diseased matter acting irritatingly on the nerves and which spreads ever further until at last the whole nervous system is attacked by it and rendered useless.

Not only does the cramp spread over the body from the point where the pain began but with it an ever increasing stiffness of muscles until at last the whole body is as stiff as a piece of wood.

The stiffening does not merely extend to the external parts but it acts internally; the action of the heart grows more feeble, the breathing becomes difficult, till at length a complete stoppage ends in death.

It not infrequently happens that only a part of the body is attacked by Tetanus just as Blood Poisoning may be confined to an individual part, simply because the inflammation is too weak to seize on the whole body.

Thus the stiffening of individual parts may be limited either to the hands or feet but especially to the neck. If the disease is of a mild kind it is possible that the cramp will spread no further but rather by degrees decrease until the pain quite goes. If on the contrary it grows worse and worse it will be at the cost of life.

To cure such a patient two things must be kept in view; first that the further progress of the tetanus must be stopped and second that every thing that has become corrupt must be made good.

As the disease began by the penetration of corrupt matter which caused inflammation, this point must be first operated upon and then the whole body, in order to dissolve and extract everything corrupt.

According to my opinion there is no remedy equal to compresses of hay-flower-decoction as warm as the patient can bear them. Begin with them first on the specially affected part and continue them as far as the disease has spread. What these Compresses have been unable to do complete by sweating whether the individual parts or the whole body; in any case the chief thing to be done is to dissolve and extract as rapidly as possible.

My brother in law wounded himself but so slightly that he was scarcely able to indicate the spot. For sometime he was quite able to use the limb but by degrees complete rigidity set in and spread to about half his body. The hands were quite stiff, the neck was so rigid that he could not move his head, and his tongue was immovable so that he could not speak a word; his mental powers on the contrary were quite in order. I made him put on a shawl dipped in hay-flower-decoction as warm as he could bear it; at the end of three quarters of an hour it was redipped in the warm decoction. This was to cause an equable warmth and so dissolve and extract.

When the shawl had been redipped the third time the invalid noticed that the pains were less and after three hours he could move his tongue sufficiently to speak. I may add that I now gave him milk boiled with fennel and a tea spoonful of anise tea as well every three quarters of an hour and it was just this last that he thought did him most good. In fact there is scarcely any plant by which one can get such excellent results in convulsive attacks as by anise.

As soon as the applications of the shawl were over I tried to operate on the whole body in order to dissolve and extract the diseased matter.

For this purpose a many folded cloth was dipped in warm hay-flower-water and laid upon the abdomen and renewed every half hour.

A heavy sweat set in and the whole body was bathed in it and the relief was immense.

This application was continued alternately with whole washings until all hardenings were dissolved and the normal condition reinstated.

. . . . .

Stammering.

One often meets people who cannot speak without stammering. It may be caused by insufficient development of the vocal organs or it may be the result of a severe illness.

It is a weary evil and often very hard to cure; the many medical attempts to cure stammering may be regarded as failures.

I am firmly persuaded that if no organic fault exists water is here the best remedy; at least it can bring a tolerable amount of relief by means of a general and local treatment of the vocal organs. Let me give an example. One day a girl, nineteen years of age, was brought to me who could not get a word out without stammering.

Naturally she was laughed at by her schoolfellows and at last she could not or would not speak at all. It was at this point that the parents brought the girl to me at Wörishofen. Fortunately no organic fault existed and therefore the stammering could only be attributed to general weakness.

The applications I ordered were quite simple viz., half baths alternately with full douches and thigh douches. For local treatment of the vocal organs I gave different teas both for gargles and for drinking; first pewter-grass and sage tea, next bark-of-oak tea, then spruce tea. I further ordered internally a cup of tormentilla and wormwood occasionally.

Walking in water and on wet stones, the first for two or three minutes the latter for ten minutes, is a most effective remedy.

As I treated this girl so I have several and in nearly every case I was successful in bringing them to a proper way of speaking.

Cramping influences on the vocal organs will often cause stammering, so will sudden fright. In either case water will remove the stammering.

I must remark here that the stammerer should speak as much as possible but very slowly and in measured time. If he does not shirk this trouble he will gradually improve his condition.

Beside stammering, the vocal organs are subject to yet another and graver evil such as cramp of the Epiglottis and paralysis of the vocal cords. By the first I understand a convulsive contraction of the vocal cords during which there is great danger of suffocation.

Paralysis of the vocal cords is inactivity of the same arising from whatever cause. I ascribe this trouble more to weakness of the system which, however. must not be taken for granted.

~~~~~~~~

Venereal Disease.

Just as good qualities may be handed down from one generation to another so may vices and diseases.

There are diseases which are not at all hereditary and concerning which we need have no fear that they will be handed on, but on the other hand there are many which are easily inherited and readily passed on.

Thus a disease prevails among men which causes unspeakable harm and passes down from one generation to another.

This disease is subdivided into three parts and according to the nature of each is called Syphilis, Chancre and Gonorrhea.

If only men would live a clean and moral life and steadily avoid immorality and vice we should not have so many victims to inherited disease.

In any one who has drawn such an evil on himself, by his own fault or has it by inheritance, the poison grows and ruins everything and every one he goes near producing one evil after another. What a fearful responsibility such an one incurs who, eaten up by syphilitic poison, transfers his misfortune to others!

How infectious and dangerous this disease is may be seen in the following example.

I knew a person who, having washed the linen of a syphilitic patient, became herself infected. The disease entirely eat away her nose and she got a quantity of ulcers all over her body although she had always led a very moral life.

If only young people would believe in the fearful results of this disease! If only they would act reasonably so that they need not waste the best years of their lives in misery!

It is not that these people alone are plunged into the deepest misery but it is a disease which descends to the third generation causing their posterity to suffer and bow down under the burden.

A man, eaten up by this poison, is not in a condition to fulfill his professional duties; he is feared and shunned everywhere and with reason. The disease ruins his body, he is plagued and pained by all sorts of troubles and finally falls a victim to it.

When once people have become miserable and infirm in mind and body through the indulgence of their passions truly repentance and wailing come often too late.

As regards the cure it is often very hard to accomplish; Many use the so called whey or quicksilver cure but neither of the three diseases with their ruinous results are really healed by it. The sufferer from one or the other can only endure a sad existence with as much patience as possible until his body succumbs.

I am convinced that, with water, improvement may be arrived at and in many cases complete cure; for the action of water is to dissolve and extract the impurities and to brace the system; of course the applications must be suitable and continuous.

It is necessary to know this disease and its origin exactly; how long it has been at work and how much bodily strength still exists before commencing the cure by water.

As in most other diseases so here there must be a general and a local treatment.

For the first, one must act on the Abdomen and choose a very simple, unexciting diet; there can be no question of alcoholic beverages, these must be completely abjured.

If the body is already much reduced begin with whole washings from bed; at first only one a day, after a few days two.

When these have been made for about six days the douches may begin in the following order; first day, a knee douche; second day, a thigh douche; third day, a half bath; fourth day again a thigh douche and so on with the addition of an upper washing.

These applications are easily borne by the patient and have a good effect.

After making them for some time a change may be made with half bath, full bath, thigh douche, and back douche. These all improve the blood, cause a strong ejection of diseased matter and heal the bodily parts attacked by the disease.

If Ulcers form one should not try to heal them up, on the contrary, one should be glad that, through them, the diseased stuff has got an outlet.

To keep these Ulcers clean poultices of pewter-grass-decoction must be laid on frequently and now and then compresses of hay-flower decoction.

The decoction of hay-flowers extracts, while that of pewter-grass cleanses and contracts. If the patient has abdominal troubles bandages of hay-flower-decoction will have a very good effect.

Internally a tea of wormwood and centaury herb can be used with much advantage to the stomach.

Angelica root is good for extracting the poison and juniper berries are good for the patient.

If this treatment is steadily pursued for some time the body becomes more braced and the best results accrue from it.

We have had many patients here of late years who had, before coming, tried every imaginable thing without success and who by our treatment have become very much better and some even cured. Of course it must be understood that after they leave us they must lead an absolutely simple and moral life.

~~~~~~~~~~~~

Hydrophobia. (Frenzy.)

During my student days I resorted regularly to an Inn on the journey from Dillingen home.

In the course of conversation the Landlord once showed me on the ceiling of the room a round hole and explained that a strong and healthy man was bitten by a dog and became mad. He further told me that the man was quite quiet at times and that he knew beforehand when an attack was coming on: he then let down a rope through this hole, having first tied it round himself, so that one could hold him down on the floor of the room above until the attack of frenzy was over.

Thus the man lived a long time until, after much suffering, he died in fearful agony.

When this disease has thoroughly taken hold of a man it is hopeless.

Those who have been bitten by a dog are in constant anxiety, fear, sadness and terror; they have a sensation of heat as if their whole body were full of fire and drinking does not help them; and then comes on an attack of frenzy which lasts till the sufferer is worn out.

This disease is usually considered incurable; I am of opinion that it has its remedy like every other non-hereditary disease.

As soon as the disease has penetrated somewhat deep into the system the patients have an intense aversion to wet and water, hence the name **hydrophobia**. All sorts of remedies are sought for at the Chemists which are as a rule quite useless.

I am convinced that this disease may be cured if it is not left too long, and I should use the same treatment as in blood poisoning; for is there any difference between blood poisoning brought on by a wound with a splinter of wood or a nail and one caused by the tooth of a mad dog?

I have cured many cases of blood-poisoning for which no one believed that a cure was possible and, indeed, the cures almost bordered on the miraculous seeing that the poison was extracted from the body as fast as a splinter or nail from the wound.

Our Ancestors tried to cure hydrophobia by cutting off the hairs of the dog which had bitten the person and laying them on the wound in the belief that they would draw the poison out.

If as I believe the thing is to extract the poison from the body, then water will be as effectual here as in other blood-poisoning.

The dog bite may end fatally so may other blood-poisoning. If the dogs hairs can extract the poison, much more quickly and certainly can water do it.

Now comes the question, How can the poison be extracted?

I will state a case; a mad dog bites some one in the foot, the inflammation spreads rapidly, great pain sets in, and heat, fear and anxiety increase to the highest degree — well, boil or soak hay-flowers as quickly as possible in boiling water and poultice the wounded limb up to above the painful part; let the poultice be as hot as possible, keep it on eight hours and renew it every hour, then the pain will diminish, the heat is quenched and a normal condition established. As soon as the cramps and pains cease healing begins.

This is the way it acts in blood-poisoning and thus it acts in madness which has arisen from the bite of a mad animal. If the pain spreads to other parts of the body all those parts must be treated in the same way and very soon a profuse sweat will be produced and by it all poison will be extracted. If the convulsive condition should return it is a sign that small fragments of the poison still remain in the body and these must be drawn out by degrees; in this case the applications must be made anew even if somewhat weaker and less frequent, in fact they should be arranged according to the nature of the relapse.

As soon as a person is attacked by hydrophobia in consequence of the bite of a mad dog, the wound should first of all be cleansed with arnica and water; then hay-flowers which have been soaked in boiling water should be placed on a coarse linen-cloth and laid on the wound as hot as possible and wrapped round.

As soon as the warmth diminishes the poultice must be renewed, the cloth is re-dipped in warm hay-flower decoction and bound on afresh and this must be continued till all pain has gone. It is even better if the bad foot enclosed in hay-flower poultice is placed in water as hot as one can bear it; in this case warm water must be added every half hour to keep up the same temperature.

The same hay-flower-poultice should not be kept on longer than two hours as the power of it would not last beyond that time. In this way all pain will have gone in from two to six hours.

As one operates on the affected parts so must one operate on the whole body because in blood poisoning the entire body suffers.

To secure this it is not necessary to wrap the whole body in a cloth soaked in hay-flower-decoction, it will suffice if the body gets into a profuse sweat or perspiration which will extract all harmful matter which has penetrated within.

This perspiration may be brought about by taking a whole washing every two hours; if this does not answer put on a shirt dipped in hay flower decoction. Should hay-flowers not be at hand use oat-straw which will not be without effect although I give the preference to hay-flowers.

I related a page or two back the story of a mad man; how ought one to have proceeded with him? I should have treated him like a cholera patient and have got him into a profuse perspiration.

If a little bag filled with hay-flowers and dipped in hot water is bound on the body as hot as possible it will produce in a quarter of an hour a profuse sweat which will break out all over the body.

In addition to this I should wash him twice a day in water and vinegar so as to strengthen him or if he could have stood a bath I would have dipped him daily in water and continued to do so as long as symptoms still showed themselves.

There are beside the bites of dogs those of cats and other creatures and I have cured many of these.

It is usually considered a good sign if blood flows out of the wound because in this case poisoning is less feared. Country folk even tear the wound to make it bleed and press it out, a plan by no means to be despised.

Even when the wound has just been cleaned with arnica tincture I advise a hay-flower bandage in order to prevent inflammation spreading. Do not put the swollen hayflowers into a bag but lay them direct on the skin for an hour and a half: if by this time the inflammation and redness have gone lay on a cloth dipped in arnica tincture; if, however, the inflammation is still severe then the hay-flower bandage must be renewed and continued till it goes.

Typhus.

Typhus, or as it is sometimes called Nerve Fever, developes various forms each with characteristic symptoms. There is for example Typhus in its mildest form, **Abdominal Typhus** and **Spotted** or **purple Typhus** so called because of the purple spots which appear on the body: this form is extremely infectious. Then there is **Relapse** or **Low Fever** which arises usually from disturbances in the blood.

In the Water-cure treatment we do not concern ourselves with the individual forms and characteristics displayed by Typhus but, on the contrary, we deal with it generally that is, go to the root of the matter. Still the greatest care and consideration must be exercised in the application of remedies. It would never do to treat Typhus lightly, for neglect or unsuitable treatment would be productive of the worst results, both to the patient and his neighbours.

The form of Typhus most general is **Abdominal Typhus.** It makes itself known by great and continuous languor, uneasy sleep, heavy head and giddiness, which, as they increase, make the patient appear as though he were drunk.

Added to these shivering fits, alternating with heat, occur and the ears tingle, in a word the patient feels obliged to be in bed whether he likes it or not. He suffers great thirst and diarrhœa.

This condition lasts five or six days.

24*

After this, the disease developes fully, and little spots become visible on the breast and abdomen. Now is the time when it is either the one or the other, death or life.

If the invalid gets into a gentle perspiration and obtains a quiet sleep and grows less thirsty there is every hope of his recovery.

If, however, the contrary obtains and a decided loss of power appears, if the patient gets delirious and the pulse weaker and more irregular then, as a rule, it is death.

The disease which throws whole districts into the greatest fright is that most dangerous form of Typhus known as Spotted Typhus or Purples. Wherever it appears it places many human lives in danger partly by infection and partly by the rapid course of the disease.

The danger of infection is so great that the slightest evaporation produces it.

I cannot here sufficiently impress on you the absolute necessity of cleanliness in these diseases, for not a few are distinctly produced by dirt. Above all, the sick room must be diligently aired, for nothing is better than fresh air.

The bed and body linen must both be scrupulously clean; it is advisable also to isolate the patient, in a word the most perfect cleanliness and punctual exactitude must be observed.

The symptoms of Eruptive Typhus are nearly the same as in Abdominal Typhus viz. lassitude, shivering alternating with heat, loss of sleep and appetite, head ache, bad stools and the like.

Naturally the patient is compelled to remain in bed. This condition goes on for some days until red spots are seen on the whole body, indeed they appear in such a mass that it looks as though the skin were powdered with them.

The condition of the patient at this point is very critical and unless a change takes place within fourteen days there is but little hope of his recovery.

The Relapse or Low Fever owes it origin mostly to bad blood.

It frequently makes its appearance in crowded sleeping places or where people do not live good lives. It differs from General Typhus of the mild kind in that it breaks out without warning and that the patients suffer violent pains in their limbs. If this form of Typhus is not treated in the right way and if the bad stuff is not extracted and the blood cleansed it may have bad results for the patient.

There is a fourth kind of Typhus which must not be overlooked and that is **Head Typhus** a most dangerous form of the disease. It shows itself in violent headache and occasional vomiting. Gradually a sickly exudation forms in the head which is very painful and in most cases there is no chance of recovery. Water, cleanliness and strict diet play a very important part in the treatment of Typhus. The great fault in the use of the water is that patients remain in it far too long. Even ten minutes I consider too long if the patient is to be strengthened.

My plan of douching is much more successful than any other method.

My great object is to stop the development of Typhus rather than give it help. If the disease begins with fever whole washings every hour are of great advantage for they produce perspiration and in this extraction begins.

In other methods baths are begun with a certain degree of warmth I, on the contrary begin even at the first sign of fever with the cold water treatment and try directly to extract the diseased matter.

I do not allow the half baths to last longer than six seconds and therefore they do not weaken the patient like baths that last ten minutes and during which he is rubbed and massaged.

I do not at all like the use of icebags; in the first place they are quite unnecessary and any help required may be quite as efficiently given by compresses on the head which should be changed every five or ten minutes.

Beside these compresses on the head I cause cold water compresses to be laid on the Abdomen for an hour and a half and renewed at the end of three quarters of an hour but only once or twice a day.

Short bandage and upper and lower compress are very beneficial in Typhus but they should only be used when quite needful. A cup of tea daily made of juniper-berries, sage and wormwood is very good for the diges-tive organs. I have cured the worst complications with these simple remedies even when heart trouble existed at the same time as the fever. The Nurse should be careful to keep the mouth of a Typhus patient clean and moist so that the tongue does not get dry.

St. Vitus' Dance.

St. Vitus' dance is a disease from which both chil-dren and adults suffer although the latter are not so subject to it as the former. In my youth I never heard of St. Vitus' dance; I can only remember one single example where a child, through a great fright, got into the condition known to us as St. Vitus' dance.

Now-a-days it is undoubtedly of frequent occur-rence.

If a child is afflicted with this disease the muscles of part or the whole of the body are active as with feverish excitement; the face is distorted, arms and hands twitch and toss with convulsive movement, the head turns now right, now left, and twists in all directions.

These movements which at first are scarcely noticeable may increase to such a pitch of violence that a strong grown up person can barely hold the child of ten or twelve years old so afflicted.

St. Vitus' Dance frequently occurs after severe illnesses as for instance after scarlet fever, diphtheria, or joint rheumatism and the like.

My opinion is that it occurs in those people whose systems are very weak and in whose bodies diseased matter remains after the illness, and I believe that they can be quickly cured of St. Vitus' Dance if an eruption can be called out on the body or even on part of it.

It is my further opinion that children of the present day are so delicately reared that they are never strong enough to eject the used up bad matter in their debilitated system.

These children generally have a sickly appearance, a bad appetite and possess no healthy normal strength: they are usually very excitable, suffer from poverty of blood, and succumb to the smallest mental exertion.

The proof is afforded by the circumstance that St. Vitus' Dance decreases and a healthy natural power is installed as soon as such children get good nourishment and are braced. Medicines are but of little use in this disease and yet if the child is left to itself to get well the disease may linger on for weeks and months, whereas it is removed in a very short time if the patient tries another climate, or gets a better diet.

I have already indicated how such invalids may be cured viz. by bracing, by wholesome strengthening diet and by dissolving and extracting the bad matter left in the system by weakness or disease and which it is absolutely needful to remove.

I knew a boy of twelve years old who had St. Vitus' Dance so badly that his father was unable to control him.

He came to me complaining bitterly at having such an unmanageable lad. I told him that if he had whipped him on the back with stinging nettles he probably would have been quite quiet.

The father thought that he certainly could do that and went straight away for a bushel of stinging nettles with which he beat his son on back and chest. The result of this whipping was that the boy came in a rash as thick and strong as that of scarlet fever and when this was cured the disease had vanished.

I have frequently put on a person afflicted with St. Vitus' Dance a shirt dipped in salt water and wrapped him round thoroughly in a blanket, and have repeated it three or four times a week. This treatment has brought out a red rash and as soon as this healed there was no more St. Vitus' Dance, a proof, if one were needed, that the excitement of the muscles arose from bad matter.

The water applications must be of the simplest and mildest character in order that the system, already excited, should not be irritated.

In order to brace the system, the walking barefoot in water or on wet stones answers best. This may be practised daily once or twice for the space of a quarter of an hour with the water reaching half way to the knee: there would be an advantage in walking barefoot; in the house also should opportunity be found.

Children, who are very weak, should be washed morning and evening in vinegar and water over the upper part of their bodies; after a few days they can quite well bear half baths which are the least exciting in their action.

At first it is sufficient to take the child from bed and dip him in water for one or two seconds then put him back directly and well cover him up.

Having continued these dippings for some days change them for, one day, a half bath and the next day a full douche.

One or two days in the week there should be no water applications; let these **rest days** be used for walking on wet stones which braces the feet and acts well on the disease.

If the child is strong physically and looks well nourished the body must surely be spongy. This being so, a shirt dipped in hay-flower-water or in vinegar and water should be put on the child once a week for three quarters of an hour or even an hour: both these applications are very effective and successful. As regards the diet, only the most nutritious food should be chosen, such as is easily digestible and able to produce good blood; on the other hand all exciting food and beverages, especially alcoholic, should be strictly avoided if one does not want to pour oil on the flames.

For breakfast, I recommend malt cooked in milk or strong broth and the food generally such as I have ordered for epileptic patients. I must say here that butter on bread and other greasy food which people give their children is of no use here; on the contrary pot-cheese with black bread is an excellent diet for children and gladly eaten by them. Do not give them much at one time rather little and often so as to avoid the danger of overloading the stomach which excites the system. Five times a day would not be too often if only a little is taken at the time.

Internally such remedies may be taken as to cause a good digestion, brace the system and cleanse the blood.

Tea of sage, ribwort, stinging nettles and strawberry leaves acts favorably on the blood.

Alternately a tea of oak bark, tormentilla and centaury herb may be taken but always in small portions such as a tea spoonful four or five times a day.

From time to time as much as will lie on the point of a knife of angelica and tormentilla powder is very good in its effect.

I recommend specially for such children burnt-bone powder, a very small quantity twice a day as much as can be placed on the point of a knife.

Burns and Frost Bites.

Burns.

It often happens that a house on fire nothwithstanding all that is done to arrest the mischief is reduced to ashes, while in another the fire is stopped and the damage can be made good.

It is just so with the human body, a burn or heat causes such severe destruction that help is useless while in other cases the mischief is less and help can still be given. As a rule if the third part of the upper surface is destroyed by a burn cure is impossible. Burns occur so often that I have had frequent opportunities of treating them and curing them.

Burns may be divided into three degrees; the first and easiest is that in which by the fire or heat redness appears which is I grant very painful but which grows more bearable as time goes on. The second is that in which blisters, large or small, form on the skin some as large as a good sized apple and which contain a watery fluid.

The third is that in which by the action of the fire or heat not only is the skin seized on and inflamed but the flesh lying under it as well, so that pieces of burnt skin and flesh fibres exist in the wound. This last is of course the most dangerous of the three degrees.

Such burns may be caused by boiling water or other hot fluid, by direct contact with fire or a fall into it; which last sometimes happens in a conflagration, as well as by other accidents too numerous to mention. In the first degree of burns which are most frequent country-people try to cure them by the simplest means: sour-crout-water is taken from the crout tub, compresses made with it, which are renewed every half hour and the burnt part is very soon healed. Others rub raw potatoes on a grater and lay them on the burnt part. By this the heat is soon removed and the ruined juices extracted; while others again take linseed oil and rub the place, or lay it on gently and so get help.

Water mixed with vinegar and put on acts well as do all cooling and extracting remedies, only care must be taken that the bandage does not remain too long on the burnt place without renewing it, otherwise the heat and naturally the pain will be increased

Healing is not so easily accomplished in the **second degree** if blisters have formed. Even for these burns, if they are not too large, Country folk use household remedies. If the burnt surface is very large it may easily become dangerous.

My opinion gained by long years of experience is that those remedies are best by which the entrance of fresh air is prevented reaching the burnt place, and by which the heat and corrupted matter are drawn out.

Take about three tea spoonsful of sour cream, stir this well, and cover the burnt place as thick as possible with it so that the air, which would increase the pain, is completely shut out. Put the plaster on as thick as possible with a clean feather, over which lay a very soft little cloth dipped in cold water and over this again a dry one which must well cover the plaster and bandage on all sides. Every now and again according to the degree of heat and pain the first wet compress is renewed; but the greatest care must be taken not to tear it off by force, it would be better to dip the affected place with its bandage in water or pour fresh water slowly on the little cloth until it is again wet through, and lay over the dry cloth as before. Twice a day the cream plaster must be renewed or rather freshly laid on.

If the blister does not break of itself and discharge then it must be opened a little on the side, so as to allow the skin to remain lying as a covering, until a new skin forms under its protection.

In this way healing generally sets in fast.

In burns of the third degree in which real burn-wounds arise act in the same way.

Put the plaster of cream on two or three times a day according as the wound is better or worse. Let the compress never get dry and at least twice a day the **wound** must be **washed clean** so that nothing hinders the cure.

For this washing use pewter-grass decoction in which a little arnica has been mixed.

In burns of the second and especially those of the third degrees the patient often feels an intense feverish heat in the **whole body** and one can with truth say that the whole body is sick and affected, therefore one must operate on the entire system.

In this case take the first day a short bandage, the next day a whole washing, and the third day a half bath, or one of the other applications, according as the burnt-wounds allow; at least one application should be given daily.

I recommend very strongly to those who have suffered from severe burns to take at the commencement a spoonful of fine salad oil. One would hardly believe what a wonderful effect this has in keeping the stomach active and the tongue clean and in keeping the heat of the body under. If the patient dislikes taking a tea spoonful at one time he may divide it into three or four portions if he takes the quantity during the day or, if he prefers it, he may take from ten to fifteen drops four times a day on sugar.

If the burns are severe and wide spread great disturbances arise both in the organs and in the blood and in many cases produce so much corruption as to prove fatal.

To combat this increasing weakness of the blood and general inactivity of the organism as well as to maintain the whole system in healthy activity it would be well to take every day a cup of tea of wormwood, sage, and ribwort or alternately with that a tea of angelica root, tormentilla and oak bark.

Frost Bites.

All who have witnessed the pain and destruction brought about by a severe burn know that the person so burnt is indeed a martyr, but no less direful in its effect is the contrary of burns viz. that of Frost Bites.

These occurred much oftener in my youth than now, principally because travelling is rendered much easier by the Railways, while at that time one had to make journeys generally on foot.

In my home-parish we had one or two or more cases of frost bites every winter and still oftener there were frost bites of individual parts such as ears, hands and feet.

Wandering over unfrequented paths through deep snow the lonely traveller, overcome by fatigue, would sit down to rest a little. Naturally he went to sleep and awoke in eternity. Yet it sometimes happened that some one came that way, found the half dead man and rescued him by quick and ready help.

The man, thus saved from death, could then tell **what happened when one froze.** Thus such a man related to me his experience; he had to travel for three hours and a half in the snow; the first hour he got on right bravely and did not freeze a bit. But soon after, the snow got much deeper and he often sank in up to his knees and grew more and more tired: an overpowering sleep attacked him so that he was unable to get on; he sat down to rest a little intending to walk on

again after a few minutes. In spite of his efforts, however, he fell asleep and woke up in his home whither
he had been taken. The person who had rendered him
help and brought him home said he found him on the
way half an hour from his house so fast asleep that he
thought he was dead. On arriving he was laid in a
trough of ice cold water and left in it for some time.
He was then taken into an unheated room and laid in
bed where he was washed frequently with cold water.
This was the story told by the man himself; he got well
again but had he slept longer on the way his recovery
would have been doubtful.

It was in this way that Country folk in my youth
treated frozen people; they were fully convinced that if
they took a frozen person, who still had life in him,
into a well heated room he would certainly die because
the transition from cold to heat was too sudden for the
system to bear. If you put an apple, which is frozen
for the first time even as hard as ice, into cold water
and let it lie there a little time the water draws the
cold out and there forms round about the apple a thick
crust as hard as the ice in the open air. Wonderful as this
seems I know it is true for I have often tried the experiment. I have also frozen the same apple a second
time and treated it as at first but it never answered to
it, on the contrary it remained frozen and spoiled. This
proved to me that the life power of the apple was destroyed and could not be renewed by the second bath; it
also made it clear to me that the cold water in which
the frozen man was dipped extracted the cold from his
body and reanimated it.

I once knew a butcher who having to go for a
head of cattle lay down on the way through fatigue
and exertion with the result that both his feet
were so frost bitten that it was thought he would
lose them. Fortunately an old peasant came to his
help — he brought into the room a tub full of snow
placed the feet into it above the frost bitten part and
began to rub them vigorously with the snow. It was

exceedingly painful to the butcher but the old peasant took no notice of this but went pitilessly on with his rubbing until the frost was all extracted from the feet and he had the use of them once again.

I knew of two other people who, in like manner, had frozen their hands, the treatment they received was that the frozen members were bound up in snow and left in it until they became well.

Frost bites occur even more frequently on the ears than on hands or feet and if snow is procurable it is rubbed gently on until the cold is all extracted.

One sure sign of frost bite on any part of the body is that it quite loses its natural colour and is as pale as death.

I now ask the question — If one can extract the cold from a frozen apple by cold water, is there a better remedy than snow or ice-cold-water to extract the cold from a frozen human body or any of its parts?

If the frozen parts are not thus treated the cold remains in them and by degrees they are destroyed, for the action of the cold is to destroy the juices and the organs, and to hinder the circulation of the blood.

Supposing any one to be suffering from frost bite he need not lose hope of life if he is at once put into quite cold water and the affected parts rubbed gently. He should then be put to bed so that the normal warmth may gradually return.

If, after some hours, he is redipped in cold water a similar process takes place in his body as in a burn; as in the one case the burning diminishes so in the other the consequences of frost bite are extracted.*)

Those parts of the body which have suffered in a special way must also be treated with special care by

---

*) Note. Instead of the dip in cold water it would be of use to make a whole washing every half hour until the return of normal warmth.

wrapping them in a cloth which has been dipped in cold water and is to be frequently renewed.

This experiment should be continued until the patient is free from the consequences of the frost bite. If however the bodily parts such as hands or feet are quite frozen then these unavoidably mortify.

Teething of Children.

The teething of children may be productive of such evil results as to cause death.

The teeth come at different times and the first teeth cause the little ones trouble.

Why? I think I am not wrong if I say that the children who suffer much with their teeth are very weak and that obstructions exist in the blood and juices. The whole body is insufficiently developed and the teeth appear with difficulty.

Those who from their earliest hours have been dipped in cold water get through the time of teething with much less difficulty: indeed I know of no child who has been brought up under the water treatment who has had trouble in teething.

If a Mother, instead of putting her little treasure into a warm bath every morning, would dip it from its bed in cold water and again put it back to bed it would be saved much trouble and sickness. One would think that the thing a mother would most desire is the health of her child; If so she must act accordingly; she must accustom her pet to bracing and strengthening of its body and to wholesome food; thus she would show true motherly love.

The blind fondness with which many Mothers surround their children has no value whatever, on the contrary they may ruin them with their stupidity.

Medicines are of no use during the teething period. The best thing is cold water which brings the circulation of the blood into proper order and strengthens and maintains the whole organism.

The children who are braced from their earliest days remain free of many troubles which less favoured children are subject to.

Gum-Boils.

A girl who had a swelling on her cheek paid very little attention to it thinking that it was not at all an unusual thing to happen. But the swelling increased, became firmer and harder and the pain so great that she could scarcely bear it.

The Doctor was called in and he pronounced it to be a Gumboil.

As it became no better and the pain grew worse it was operated on and some teeth were drawn out. An indiarubber tube was inserted to allow the matter to run out properly.

The girl wore this tube for more than half a year yet the pain remained, so did the swelling, and the extraction of the teeth had not helped the girl at all.

In the course of time the whole cheek sank in from the ear to the temple. As nothing that had been tried had proved of the least service, and she heard now that the Doctor was about to remove the jaw, she determined to come to me at Wörishofen.

The poor thing bewailed her pain greatly; I required that the little tube should be removed but she would not hear of it for it had been impressed upon her that its removal would prevent the matter from flowing out and so cause her death.

I explained that I would make no attempt to cure her unless the tube were removed and it therefore was taken away.

I began by cleansing the gums thoroughly every hour with pewter-grass-water and making her gargle her throat often.

The cheeks were douched twice a day from the hair to the chin strongly with cold water. At the end of a fortnight the discharge of matter was certainly less but the cheeks remained sunken. The pains however decreased and at last vanished altogether. At the end of eight weeks the cheeks began to get fuller and the discharge grew gradually less till it stopped entirely and the girl was cured.

While with us the girl was treated properly with water applications so that all that had become diseased in her system through the gums was removed and the whole body got into a better condition.

The girl got a fresh healthy colour in her cheeks and at length she said "I know now that I never could have been really healthy for I feel so different."

Diabetes.

This disease appears much oftener now than formerly: it is an ejection of sugar into the Urine by which the body becomes greatly emaciated and weakened if not suppressed at the right time and a change made in the mode of life.

Great beer drinkers and people who eat largely of food containing sugar such as confectionery, sweetmeats and such like are generally subject to this disease. Corpulent people and those who lead a sedentary life are also liable to it. The first sign of the disease is an inordinate thirst, dryness of the mouth, frequent passing of urine and an increasing emaciation of the body. However unimportant it may seem to many people, yet this disease unless checked may cause the whole body to decay.

It lasts on an average three years, in rare cases eight years, and the cause of death is mostly due to extreme emaciation with the addition of lung troubles and exhaustion.

Many people suffer from a collection of sugar in the Urine which although not serious demands care.

It is a disease which is easier to heal in the first stage than when it has made great progress.

There have been patients here whose urine contained ten to twelve per cent of sugar and who, by a simple mode of life and suitable cold water applications, were completely healed without any relapse. When a person observes that symptoms of this trouble appear he should at once have the urine examined and if sugar be found in it it should be analysed from time to time to see if the trouble grows greater or less.

In medicine much weight is laid on keeping to a prescribed diet. Doctors allow meat, bouillon, eggs, cheese, butter, bacon, fish, red wine and so on, while they forbid sweet wines, sweet fruits and such like.

In many cases this is quite right yet I maintain, after many years experience, that so much anxiety about the food is unnecessary. Just as in other sicknesses I recommend here simple, non exciting food; in a word, I advise an easily digested yet mixed diet which can supply the stomach with many nutritive ingredients, and afford the body strength to resist evil matter.

Beer, alcoholic beverages, sweetmeats and a diet which tickles the palate do not suit at all. They act like poison on people suffering from **diabetes.**

Rather let them eat strong soup and green vegetables alternately with farinaceous food. If the mode of life be regular and injudicious food avoided, the condition of the patient will improve, always providing that water applications are taken so as to strengthen the decayed body and brace it.

The best way of proceeding is, — during the week to take two or three sitting baths, one or two half baths,

25*

a back douche and, if the patient has a tolerable amount of normal power, a full bath.

For internal treatment simple remedies should be employed for the improvement and strengthening of the stomach and among these there is none better than tea of centaury-herb, sage and wormwood of which a table-spoonful every hour should be taken; for quenching the intense thirst tormentilla tea is best.

Beer, wine and such like drinks should be strictly avoided unless in the case of a person who has always been accustomed to take a daily portion of wine; then I might be induced to allow him a couple of glasses but the quicker he learns to do without it the better.

With these few words I have said all that is important on this trouble.

Those accustomed to take simple, nourishing food and who lead healthy bracing lives will escape many evils.

If, however, Diabetes has made its way into the system get rid of it by water, and to see the wonderful effect of water read the following.

A gentleman from some place in Swabia came to Wörishofen for the cure; his urine contained from six to seven per cent of sugar: he did not at first suspect his condition but as soon as he recognised it he practised the cure most energetically and was completely cured.

Another man from Straubing, who came here in an equally weakened state, brought on by this disease, became perfectly cured by the applications of water prescribed by me.

I might cite a number of such examples which prove the cure of Diabetes by water.

Eighth Part.

Miscellaneous.

On Blood-Letting. (Bleeding, Cupping.)

Some forty or fifty years ago blood-letting was so much the fashion that there were but few people who had not from time to time submitted to it. At every process some eight or ten ounces of blood were extracted from the system and this process was repeated three or four times a year, and not infrequently eight times.

One of my predecessors was bled four times every year losing eight ounces each time consequently thirty-two ounces a year and at last he died of dropsy.

The system creates no more blood than is necessary for it so that when one talks of too great fullness of blood it is simply that the circulation is faulty, causing a superfluity of blood in some parts and a deficiency in others. He, who has too much blood in his head, has generally cold feet.

It is true that after each blood-letting blood soon forms again, yet it is much more watery than that taken away and naturally the system gets weaker. At the time I speak of, it did not matter what the disease was: the Doctors, without exception, prescribed letting of blood believing that they would gain help by it.

Of course there may be cases, even now, when blood-letting is indisputably necessary but they are very rare. Any one who understands how to rule the circulation of the blood by cold water will never require to let blood. Cupping was also frequently made use of half a century ago and was indeed so fashionable a practice that nearly every one during a year allowed ten or twelve cupping glasses to be applied to him.

I could not understand how people could be so stupid as to let their backs be so chopped up: Thus in inflammation of the lungs one always resorted to cupping, equally so in case of paralysis. In spite of the very few good results attained by this method, it was long before the conclusion was arrived at that this treatment was very prejudicial to the system. So weakening was the process that it not infrequently happened that during the blood-letting the patient gave up the ghost. Now-a-days one is tolerably free of this nonsense yet there are still some few who adhere to the folly. In order to get strong, healthy blood and a proper circulation no medicines or foolish practises are required, cold water will do everything for us and in the right way.

As regards the increase and improvement of the blood by means of the far famed preparations of iron a sensible Doctor has assured me that they are utterly useless. This is also quite clear to me; a healthy man and especially one who knows how to use water properly needs no preparations of iron and the sick man, who does not choose a simple diet and who will not practise bracing, may take as many preparations of iron as he likes but they will never provide him with wholesome blood.

Ice-Bags.

Icebags are readily made use of in ruptures, tumours, inflammation of the lungs, pleurisy and contusions. I completely reject and disapprove of them as a remedy; a decision arrived at after many years experience.

In very many diseases it is necessary to extract the heat but this cannot be done by means of icebags since they do not extract the heat but merely concentrate it on one spot and create blood obstructions.

If instead of applying icebags one would wrap the feet in a wet cloth or give the patient a knee douche or a shawl then the blood would be led downwards and the heat arrested. Finally I refer you to my Will p. 199.

Vomiting.

Emetics used often to be taken and there are people who now make a practice of purging their system twice a year by purgatives. I do not approve of this proceeding because it notably weakens the system.

We are sometimes obliged so to operate that harmful matter is removed from the stomach as fast as possible by vomiting; this is the case after the swallowing of poisons such as belladonna, poisonous fungi and the like.

It often happens that children eat deadly night shade because they think the berries are good to eat, the bad result of such a mistake is soon seen by the enlargement of the pupils, vomiting, heavy head, giddiness and so on. How is first aid to be given to a child who has thus eaten of the poisoned berries?

First give a good quantity of warm bark of oak tea by which the poisonous stuff is made indissoluble and removed by vomiting. This tea is preferable to warm milk.

As soon as the child has vomited use cold douches, both full and back douches, and put the child to bed.

On Aperients, Clysters and Mineral Waters.

The use of Aperients was quite as much the fashion forty or fifty years ago as blood-letting and to some extent it is still so.

These aperients are largely made up of such things as have a prejudicial effect on the system. With such aperients a violent attack is made on the stomach forcing it to give up all. This naturally leaves a certain weakness behind. I know that patients are recommended to take many sorts of pills, salts and mineral waters. I cannot give one of them a good word having received enough of them myself from the Doctors: I reject all these remedies except in cases of necessity because with water the action of the bowels can be so regulated that no purgative is required. It is true that in my books the Wühlhuber pills are certainly mentioned but they should only be used in the greatest emergencies; besides, the system should never accustom itself either to pills or any other aperient.

I have arrived at the conclusion, after long years, that a spoonful of fresh cold water has an extraordinary effect.

A hydropathic Doctor told me that he knew of more than five hundred people upon whom spoonsful of water had quite a favourable effect.

People have arrived here who have had no action of the bowels for eight or even twelve days unless they took the strongest aperients. We allowed these nothing more than a spoonful of water every hour and most of them after a short time were quite rid of their trouble.

Many of them improved in five or six days while others did not get into order for five or six weeks. In very obstinate cases I made the patient sit every day on a wet cloth which also had a great effect on the action of the bowels.

Many people are of opinion that half a pint of water would act better and quicker than a teaspoonful — Buf my dear Readers, this is not so. Merely the spoonful o-water taken punctually every hour will produce the der sired effect.

Again people too readily use Clysters or injections. These I consider wrong as by every injection an act of violence is done to the rectum: therefore they should only be used in extreme emergencies.

The too frequent use of salt and Mineral Waters is to be condemned for they weaken the organs. Every butcher can tell us which farmer has used most salt in feeding his beasts by the amount in the bowels and entrails of the slaughtered animals.

Those who have been fed with much salt have such rotten entrails that they cannot be made use of in the making of sausages.

I again repeat, because of my firm conviction, that the human system is greatly harmed by drinking Mineral Waters.

If we enquire into the consequences of drinking Mineral Waters, one hears "the first time it did me good, the second less good and the third time I got no good at all."

If in diseases where great heat exists, a rapid action of the bowels is necessary take a salt spoonful of Aloe powder cooked in water.

This beverage acts gently so also does tea of briar blossom taken in small doses but of all these I give the preference to a spoonful of water every hour.

The sick people who are here for cure always keep a glass of water beside them from which they conscientiously take their spoonful every hour. Many, who at first could not believe in its efficacy, have been completely convinced of its power.

Loam. (Fuller's Earth.)

In my childhood I often noticed that when disease or accident occurred in any member of their families or among their cattle the Country folk beat up loam and laid it as a plaster on the bad part. Instead of water they mixed it with Vinegar. If for instance one of the Cattle got a blow or knock on the foot a loam plaster was the simplest remedy and the first made use of. This was done to oxen and horses alike. If a horse had a swelling anywhere which was very heated a loam plaster was laid on and the heat was quickly removed. It stands to reason that what gave help to the domestic animals might also have healing power for human beings. My experience has proved that this is so — the evil effects of a blow or a bruise are equally set aside on man's body by the loam and if applied to tumours it removes the heat effectually.

As I learned to know herbs and their effect better, my experience of the benefit of loam widened and I found that many bodily hurts and many evils were more quickly and easily healed by the application of loam than by any other remedy.

A servant, whose knee was much swollen, could not tell whether he had received a blow on it or whether he had sprained it by a false step.

He had used many remedies but without much success seeing that the heat returned very quickly. I advised him to beat up loam into a fine paste, spread it thickly on linen and bind it on very closely: as soon as the loam got dry or the heat increased the poultice was renewed. In about three days the knee was so much better that he could walk well and do his work.

All know how rapidly the sting of a hornet produces a large swelling, vomiting and blood poisoning; one of my comrades got just such a sting on his cheek; the whole head swelled with great pain and one eye was so swollen that he could not see at all with it; he had loam beaten up well with vinegar, spread it on the face and renewed it every hour. The heat decreased, so did the pain and in a short time the mischief was wholly removed.

Country folk use loam in this way for the stings of bees and wasps and with the same results.

As time went I made many experiments with loam on creatures.

Hundreds of cows are ruined every year by what is known as Milk fever; the farmer knows it well.

By way of experiment I rubbed beaten-up-loam over the backs of several cows who were suffering from this disease and then covered the creatures well up with a blanket.

Instead of making fresh poultices I caused the cows to be douched with cold water and vinegar as soon as the loam was dry; then the heat quickly subsided and in a few hours the creatures were saved. This has not only happened once or twice but very often.

Loam can be used on the human body with great advantage in high fever. In this case it is not necessary to make a thick poultice of the loam, but beat it up into a rather thick solution and into it dip a cloth and when quite wet lay it on the body and repeat this when heat sets in again or when the cloth is dry. If a person suffers from **nervous palpitation of the heart** there is no better or quicker remedy than a compress of loam water on the region of the heart. It must however be noticed that the compress must not remain on longer than ten minutes or a quarter of an hour unless it be re-dipped. A person who suffers from a pressure of blood to the head should, on retiring to rest, wrap the calf of the leg from the ankle to the knee with a single cloth dipped in loam water.

By this simple remedy the blood is extracted from the head and the ache ceases. The cloth may remain on for two or three hours or even longer because in this case the blood is directed or rather conducted gradually downwards by the developed warmth, whereas in nervous palpitations the loam compress must conduct the blood away from the heart and therefore should not get hot.

If the body is distended and filled with gas a four fold cloth should be dipped in loam water and laid on the abdomen and renewed after half an hour or three quarters. In herpes which has spread over the whole body and in which one scarcely knows how to remove the glowing heat and unbearable irritation, I have dipped a shirt in loam water put it on and over it wrapped the invalid in a blanket.

In an hour the loam-shirt has absorbed and removed very much of the diseased matter.

Many invalids to whom I have recommended a shirt dipped in loam or hay-flower-water have tried both but generally prefer the first because it acts more powerfully. The shirt dipped in loam water should however not be taken too often, at most once or twice a week, because it attacks the organism somewhat strongly.

If the herpes appear only on an individual part of the body the repeated application of a cloth which has been dipped in loam water has a very good effect. It is especially good in Lupus when the loam should be beaten up into a fine pulp and rubbed gently on; the loam absorbs, contracts and at the same time heals.

In Ascites*) a loam bandage from the arms to the knees has an equally good result. If the Ascitic patient cannot otherwise perspire and if he is still robust a loam shirt or bandage will extract a good deal of fluid.

The neck-bandage can be made either with pure water or with hay flowers or with pewter grass. The most effective, however, is a bandage dipped in loam water.

As a rule bandages and compresses of loam water are preferred in many cases to those of plain water and herbs: if, however, the mischief is deep rooted applications of loam alone are not sufficient they must be used alternately with those of hay-flowers.

_____

*) N o t e. A sort of dropsy; a swelling of the Abdomen.

Just as one can combine with water, herbs, oat straw and pine bark one can mix herbs with loam and so make the action stronger.

Our forefathers not only mixed loam with water but also with vinegar when they desired to increase the effect.

One can therefore prepare a loam pulp or poultice in which herbs are mixed or beat up the loam very fine with herbal decoctions and there is no remedy better for arresting heat, dissolving, cleansing and healing than this mixture of loam and herbs.

I myself have used loam with pewter-grass, tormentilla, marigolds, coltsfoot, ribwort and other herbs with good results.

For a long time I have made the dried loam into a fine powder and strewn it on the injured part, such as for instance in Herpes, Lupus and other sores.

Such powder absorbs the foul bad matter and contracts and heals.

People sometimes think that for an application of loam it is only needful to fetch a lump of fresh loam and lay it on, this is quite a mistake, it needs a great deal of trouble to make a delicate loam salve: First, all coarse sand and little stones must be removed and the loam must be pounded fine in a mortar ; loam may be dried on the hearth or on a hot plate ; this being done remove every particle of coarseness from it so that only the finest dust is left, out of which the loam-salve or loam-water is prepared.

In applications of loam great care must be taken to avoid too much as well as too little.

In compresses of loam, especially, one should take care that the heat is not allowed to increase and therefore as soon as the compress gets dry it must be renewed or redamped. It is just the same with a cloth dipped in loam water.

A Chapter on Teas.

Even in old fashioned Medical Science teas were used and recommended as remedies.

Those who belonged to the category of Country Doctors had great knowledge of herbs and roots and how to use them in various diseases with good results.

Gradually as Medical Science widened and changed, the simple practical Country Doctors got pushed on one side and the curative herbs were set aside to make room for other remedies. In the present day it is only here and there in some old farm house that you find herbs used as medicines.

In my opinion it was a most unwise act to banish these divine, curative herbs which have done such splendid acts of cure. Many old women who were afflicted with pains in the feet procured from God's Garden sorrel leaves which gave them relief; and every one knew the plants which were curative in diseases of all kinds.

Now-a-days the knowledge of medicinal herbs is almost completely a thing of the past.

During the many years in which I have been occupied with the water cure I have greatly concerned myself also with herbs: I have tested them and used them and must openly admit that I have obtained the best results with them.

I wish, dear Readers, that I could sufficiently influence you so as to induce you to go out into God's Garden and collect the gifts which our loving Father has bestowed upon us so abundantly; He has cared for us so wisely in his Creation as to make every little herb of use to us. If only people were not so foolish as to trample them under foot instead of thankfully turning them to account.

Only look at **Pewter-Grass**! It grows quite unnoticed on the edge of the wood and yet it is an excellent herb. There are a number of herbs which though unknown to us have great power to help us. For many years I have used herbs for teas, for compresses and baths; taken as teas herbs have an excellent effect and the preparation of them is quite simple. You take a vessel, earthen or porcelain, put in as much of the herb you require as you can take up with three fingers and pour about two cupsful of hot boiling water on it and let it stand for a little time; then strain it off and drink the decoction warm or cold according to circumstances: I must observe here that this process is not to be used with all sorts of herb-tea, some of them require greater attention as we shall see when speaking of individual kinds.

I consider it good to notice the preparation and effect of the most useful teas because mistakes are often made. Many people think they do right in swallowing whole pans of tea at once, which is entirely wrong.

A plant which is not at all liked because of its bitterness is **wormwood:** but it is of great value to Gastric patients because it creates both good appetite and good digestion. It is good for every one in short for it keeps the stomach in such order that disturbances cannot get entrance; a cup daily may be taken at one time or a spoonful every two hours.

We have now noticed two very important plants Pewter-Grass and Wormwood.

If we get further into God's Garden we find the Stinging Nettle one of the most dreaded of plants: yet the Good God did not create it for nothing; on the contrary it is of much value. Tea of fresh stinging nettles is excellent for the juices: the root of the Nettle may be used as well. Stinging Nettle cooked as soup or Vegetables·is exceedingly nutritive and sanitary. You make Nettle tea in the same way as Wormwood tea.

Look a little further and you will find quite an insignificant plant on the ground called **ribwort.**

How many a person this little plant has helped and restored to health! I therefore cannot sufficiently recommend

you to pluck it and carry it home, there is no doubt of
its value. People with weak lungs should drink this
tea constantly, and ill or well a little of it occasionally will be of benefit, for it is excellent for cleansing
the lungs.

Again, pure ribwort juice is much to be recommended. Perhaps you will ask how can one get pure
ribwort-juice?

Quite simply in the following way — Procure a
certain portion of ribwort leaves, press the juice well
out and then cook it adding to it a corresponding quantity of pure honey, let the mixture slowly steam until it
gets to a thickish consistency and admits of being drawn
off. It is filtered in a warm condition and allowed to
get cold. With this I have attained the greatest results.

Some years ago, a man from Vienna came to me in
a pitiable condition. According to the report of many
doctors his lungs were very seriously affected. He was
extremely emaciated and quite powerless and came to
Wörishofen to get help. I ordered him to take upper
body washings alternately with half baths and to take
them regularly also to take daily a spoonful of ribwort-
juice. By degrees his condition improved so much that
he was again able to follow his profession with ease.

We have said enough of this plant and now pass
on to a pretty field flower the **Centaury-herb**. If we
walk in the country during July and August we see a
mass of such flowers in many places. Centaury-herb-tea
is similar to Wormwood-tea. On account of its bitterness it is a good gastric remedyand acts well on the
gastric juices.

Every one knows dwarf-elder, sometimes called dane-
wort; it is principally used in the preparation of tea.
Tea of dwarf elder root is especially to be recommended
to people suffering from dropsy, though if the disease
is of long standing other herbs must be mixed with it,
in order to hasten the dispersion of the water. In this

case mix rosemary and juniper-berries with the dwarf-elder or, instead of the juniper-berries, pewter-grass.

More effective even than this mixture would be rosemary wine and wine of squills which may be obtained at every Chemist's. Of the latter take a spoonful every hour. If you take wine of squills let it be ten or twelve drops in a spoonful of water. A well cooked **elderberry jam** is a good remedy for dropsy. Any further information about the treatment of dropsy may be found in "My Will". But I should like, seeing that opportunity offers, to deliver a solemn warning to all.

Mankind is itself to blame for many of the diseases it suffers, partly by the mode of life and partly by the diet; by the mode of life because by it people weaken and ruin themselves taking for their guide the fashionable journals. Not only is this so among the women, the modern men are not far behind. The consequence is that mankind is thoroughly nervous and weakened from head to foot. As regards diet, people act very foolishly. Modern people of fashion instead of taking simple, wholesome food satisfy their hunger at confectioners, never thinking how they injure their systems. Again mankind will not control its passions. How many hundreds and thousands prepare an early grave for themselves by immoderate drinking. What class of people are the readiest victims to dropsy? That which indulges in a passion for beer and alcoholic drinks; but no one believes it, until he or she is chained by the disease; then remorse comes but it is usually too late.

Therefore, dear Readers, I impress on you lead a sensible life and if you do not know what that means read "Thus shalt thou live" by the old Pastor and all will be made clear to you.

Now I must return to the plants and seek for others in God's Garden. Here we see one called **Eyebright.** It does honour to its name for by its curative power it has given much help to those who suffer with their eyes.

26*

Some one asks how is this plant used? Well take as much chopped up eyebright as will fill half a small saltspoon and pour half a pint of hot water over it, let it cook some minutes and allow it to get cold and then strain it into a vessel, or better filter it through so that the eyebright is quite pure and clear.

How to make eye-baths or compresses for the eyes with this decoction you will learn in "My Will".

Other Eye-waters may be prepared in the same way for example **Wormwood-Fennel** or **Aloe**.

Briar-hip blossoms mixed with Eyebright form a good tea for constipation and cause a gentle action of the bowels.

Now we come to **Oat-Straw** — Some may deride and say "I suppose at last we shall be made to eat even oat straw at the old Pastor's."

No. I do not mean you to do that.

It is true, however, that oat-straw has a great effect on the human system: of course those who do not know its value are not the people to treasure it.

It would be well for every one if they would drink Oat-Straw tea five or six times in a month, it would be a protection against many infirmities. Tea of hips and oat-straw is good for Kidney sufferers.

Compresses of warm oat straw decoction are also of great effect in Abdominal troubles, Colic and such like.

As regards the individual application of Oat-straw such as bandages every thing needful is set down in detail in "My Will".

As there are in these days so many Gastric sufferers in the World we will now look for some herbs which are curative and helpful in this special trouble.

Of these we can find a large quantity but we will only mention a few and begin with **Juniper-berries**. The sufferer may take on the first day five berries, taking an extra one on each following day until he has reached fifteen, then work backwards till you get to five berries.

Centaury-herb, water trefoil, wormwood, angelica, fennel, anise, sweet calamus root, cummin, pumpkin seeds, gentian, rhubarb, peppermint, marjoram and thyme are all excellent remedies for gastric troubles.

Yes, the Good God has so cared for us that we have only to choose from His garden that which is useful for us.

If a person suffers from flatulence a tea of fennel will soon remove his trouble.

Many have come to me who have suffered for years from weak stomach who could scarcely touch food and who had tried every possible remedy but quite unavailingly. It is generally the case that when every medical help has failed they come to Wörishofen to the old Pastor who gives them nothing but cold water and herbs, and I must say that they generally perform a cure.

I always regret, when I see the power of water and herbs, that the present generation has so thoroughly set them aside using any trash that falls in their way instead.

I ask, what good can Mineral Waters do the stomach? None whatever, in my opinion; they only irritate the system.

For those who complain of bad appetite there is no better remedy than the water cure, a regular mode of life and a few herbs. We also specially recommend tea of wormwood and bitter clover, or angelica and wormwood, or gentian, or tea of centaury herb with water trefoil and wormwood.

There would be no end to, it if one were to relate all the wonderful effect of herbs.

We all know the trouble it is when worms penetrate into the stomach and only a few know that there is no better remedy than pumpkin seeds. About twenty to twenty-five pieces of crushed pumpkin seeds mixed with wormwood make a good tea for those who suffer from worms. After taking it for some days the worms take their departure for they greatly dislike the bitterness, and for every possible gastric trouble we have herbs to cure.

Now what is Fenu Greek?

It is the seed of the plant of the same name. One might really write whole books on this tea. Those who know the water cure are no strangers to Fenu Greek. Tea made of it is specially adapted for dissolving obstinate congestions.

Fenu Greek is also very good as a gargle for sore throats just as sage and pewter-grass tea is. Fenu Greek is not merely suitable for tea but also for plaster to dissolve swellings. In case of impending operation it is good to lay on a plaster of fenu greek for some days previously so that the operation may be more easily undertaken. I could give many examples of the good effect of this plant and have often mentioned them in my reports as well as in my books.

We have now spoken of very many healing plants and to describe all would be impossible.

Some herbs like the Carline thistle, knot-grass, pewter-grass and others I have already fully described in this book and have given directions how to prepare the teas. It must not be forgotten that when roots and blossoms are both used the former will require more cooking than the blossoms.

Many people are of opinion that the teas must be made strong in order to take effect; this is a mistake as teas so prepared only irritate. As much of the herb as can be grasped with three fingers will make two cupsful, only in the case of camomile and ribwort a little more may be used.

I will now bring forward a few which have not yet been noticed.

Rhubarb and **Fennel** make a good tea for gastric sufferers.

Arbutus-berries, oat-straw, brier-hips and pewter-grass are excellent for diseases of the bladder; the two first may be used alone, the latter are better mixed.

Knot Grass is a good tea for getting rid of uric acids.

Althea*) is good for dissolving congestions.

Tea of fir cones is specially good for bracing the throa-organs.

Bilberries, cooked in red wine, are good for severe diarrhoea.

Oak bark, cooked in red wine, is also good for diarrhoea.

Anserine gives help in convulsions; tea is prepared of it and drunk warm and it is also used as a compress. The two applications should be made at once.

Camomile poultices are good also in this trouble.

Pot-cheese although no tea I must notice here because poultices of it are so good in inflammation.

Herbs, beside being made into teas, may also be used as tinctures which are prepared in quite a simple way.

Put the chopped up herbs into pure alcohol and after it has stood some time filter it and the tincture is ready.

One can also reduce the herbs to powder but the process is difficult and I do not think it could be done at home, and it is not needful as the tea alone suffices.

There are some occasions, however, when either the tincture or powder must be used, for example, in distention of the stomach, where fluid should be avoided. The sufferers should take either gentian, bark of oak or wormwood powder in wafers.

---

*) Marsh Mallow.

Tinctures create warmth in the system but they must be used with great moderation, neither too much nor too often.

I think I have summed up the most important information on herbs. I could wish that every housewife or father of a family had a herbal medicine chest. In daily life so many cases of illness and accident occur which could be much helped by these simple herbs and water.

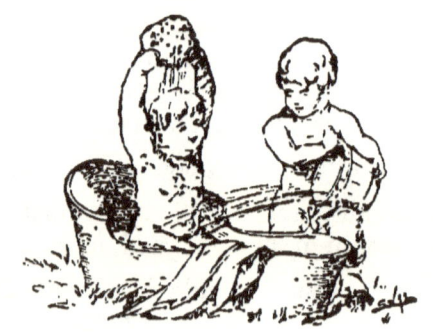

Contents.

Third Part.

Gymnastic Exercises. (Home Calisthenics.)

Fourth Part.

Formation and Care of the Human Body.

Alphabetical Index.

TABLE OF CONTENTS:

Health-Resort

recommended by the

Rev. Sebast. Kneipp:

(See Preface of "My Water-Cure", 26th Edit.)

Dr. Loeser's

Hydropathic Establishment

at Veitshöchheim near Würzburg (Bavaria)

attends its Clients exclusively upon

Rev. Kneipp's System.

Invalids of all kinds find by Father Kneipp's Cure, if not complete recovery at least marked improvement. Superior Board and Lodging in strict conformity with the ruling principle.

Comfortable Baths. Bare-foot Walkers will enjoy the grassy tracts along the river (Maine).

The Establishment is lodged in the so-called "House of Knights" (Cavalier-Bau) adjoining the beautiful Court-Garden (200 Acres) laid out in rococo style and famous for its Pleasure-Grounds, Walks, Pine- & Leaf-Wood Avenues, enlivened by Musical Performances several times a week. Excursions to Würzburg by Train (15 min.) by Steam-Boat on the Maine and by drive. Picturesque Environs. Private Lodgings to suit all tastes. Hotels.

Dining-Rooms at the New Court-Garden Restaurant.

Moderate Prices throughout.

For Prospectus and any further Information apply to the Proprietor and Manager of the Etahlishment.

JOS. KÖSEL, Publisher, KEMPTEN (Bavaria).

"Thus shalt thou live!"

Hints and Advice for Healthy and Sick People, suggesting a plain, rational Mode of Living and a natural Method of Curing by the very Rev. SEB. KNEIPP.

I. Elegant Edition. Price, Cloth $ 1.65.

II. Popular Edition for America. Price in Paper binding, net 50 cts., cloth, net 75 cts.

TABLE OF CONTENTS:

Introduction and Preface. FIRST PART: Preliminary Conditions of Health and Means of Preserving it. Chapt. I: Influence of the Light on the Health of Body and Mind. Chapt. II: Air, its relation to Health. Chapt. III: Warmth and Cold in their relation to Health. Chapt. IV: Clothing. Feet protected from the Cold. Foolish Fashions. Protection from Heat. Chapt. V: Labour, Motion and Rest. Walks, bodily Exercise, Gymnastics at home. Water as a means of preserving physical strength. Chapt. VI: Lodgings, Sick-Rooms. Chapt. VII: On Food. 1, Aliments: a, Aliments rich in Nitrogen. b, Aliments poor in Nitrogen. c, Aliments destitute of Nitrogen. 2, Drinks. 3, Salt. 4, Mineral Water. Chapt. VIII: On Meals: Breakfast. Second Repast. Dinner (at noon). Supper. Drinking at meal-time. Frugality. How many times must we eat? Chapt. IX: Education: Parental Duties in general. Parental Duties in particular. Children's Skin. Preservation of Children's Clothing. Airing, especially that of Bed-Rooms. Motion. Chapt. X: School and Profession. The Child's first School. The Child's second School. School of the Growing up. Choice of Profession. High Schools. Pupils in a Training-College. Fare in a Tr. C. Girl's Colleges. Hygiene in Girl's Boarding-Schools by means of Water. Convent-Life. SUPPLEMENT: 1, On Smoking. 2, On Colds. 3, Water resorted to by Aged People. 4, Vinegar. 5, "Toppen" Cheese.

SECOND PART: My system of Curing as taught by Experience! Asthma. The eye. General observations on weak eyes and their cure. Peritonitis. Its consequences. Caries. Incontinence. Cystorrhea. The Blood. Importance of a regular Circulation. Anæmia. Vomiting of Blood on coughing. Vomiting of Blood from the Stomach. Stagnant Blood. Putrified Blood. Loss of Blood, its consequences. Pleurisy. Its consequences. Diseases of the Chest. Emphysema. Inflammation uncured. Epilepsy. Morbid Obesity. Distortion through Lacing. Foot-Scab. Foot-Ailments. Perspiration. Diseases of the Ear. Abscesses. White Swelling. Gout. Disease of the Joints. Sore Throat. Urinary Disorder. Cutaneous eruptions. Measles. Scarlet Fever, etc. Dislocation of the hip. Infants' Diseases. Head-Affections. Spasms. Liver-Complaints. Pulmonary Diseases (Tendency to Consumption, Cold in the Head, Influenza, Emphysema, Congestion etc.) Stomach-Complaints (Dysentery, Constipation, Eructation, Dyspepsia etc.) Marasmus. Nervous Diseases. Nephritic Pain. Rheumatism, and the like. Rheumatism combined with Gout. Spinal Diseases. Tabes. Apoplectic Attack. Scrofula. Suffering from Stone. Typhus and Typhoid Fever. Abdominal Diseases (Inflammation, Spasms, Weakness etc.) Chorea and similar affections. Wrong Nutrition, its consequences. General Congestion. Wounds. Poisoning. Plunge-Bath, involuntary, how to behave after. Dropsy. The Body ruined by Dissoluteness. SUPPLEMENT. 1, On Arnica (Arnica montana L.) 2, Anæmia. 3, Gout. 4, A Word on Meal-Broth. 5, Effects of Water. a, Washing, b, Wrapping, c, Showers, d, Baths. CONCLUSION.

C. Commichau & Co.,

Solid Danish Manufacture
Under-clothing of warranted genuine Linen-Tricot

for

Gentlemen, Ladies and children.

Cure-Articles:

Bandages, Wrappers, Coverings, Spanish Mantles etc. etc. from the firmest linen, spun and woven by hand

Precisely according to precept of Rev.

Monsignor Sebastian Kneipp

made by

C. Commichau & Co., Silkeborg, Denmark.

—— Prospectus gratis and post-free. ——

Only en gros. 　　　　Agents requested.

MY WILL;

A Legacy to the Healthy and Sick.

With Directions for the correct Application of the Water Cure at Home by FATHER KNEIPP.

With Twenty-nine Photographs taken from Life, and Numerous Illustrations in the Text.

I. **Elegant Edition.** Price, Cloth $ 1.65.

II. **Popular Edition for America.** Price in Paper binding, net 50 cts., cloth, net 75 cts.

CONTENTS.

FATHER KNEIPP'S
PLANT-ATLAS

describing and picturing most accurately all Medicinal Plants mentioned in his Books, with addition of several others frequently resorted to by (country) people.

~~~~~~

In order to comply with many wishes the Editor determined upon publishing a minute pictorial representation together with an elucidating description of all those Medicinal Plants that are mentioned in Rev. Seb. Kneipp's Books. The "Plant-Atlas", now complete, is carrying out that plan. Anyone is enabled by simply consulting the „Plant-Atlas" to find out himself whatever herb he will look for in woods or fields and thus to make up in a most pleasant way that „Family-Medicine-Chest", recommended by Kneipp.

The "Plant-Atlas" is of the same Size as Rev. Kneipp's Books and has been published in 2 Editions, illustrated by Phototypes, either with plain or Coloured Plates.

a) **Edition I** containing 20 uncoloured Plates, several plants on a page, the pictures in highly finished Phototypic Printing and reduced. Price $ 1.50.

b) **Edition II** containing 41 coloured Plates; one Plant on a Page. Coloured phototypies, true to Nature. Price bound net $ 3.—.

The Plates are accompanied by an explanatory Description supplied by a distinguished Botanist and dwelling on any point of importance, for instance upon the General and Special Characteristics of every Plant, its flowering time, use, occurrence, diagnostics, mode of acting, healing power etc. Our Coloured-Phototype-Edition having met with general approval — several thousand Copies having already been sold — is already published in a Bohemian, Polish, French, Spanish, Hungarian and Dutch Edition.

JOS. KÖSEL, Publisher, KEMPTEN (Bavaria).

The Care of Children

in Sickness and in Health

by

Monsignore Sebastian Kneipp,

Privy Chamberlain to the Pope and Pastor of Wörishofen

I. Elegant Edition. Price Cloth $ 1.20.
II. Popular Edition for America. Price in Paper binding,
net 50 tcs., cloth, net 75 cts.

In this little work Father Kneipp sets forth the happiness, responsahilitie and duties of Mother hood and he instructs Mothers how to regulate their lives and how best to hring up their children. Father Kneipp also gives simple directions for dealing with the usual diseases of Children.

Table of Contents:

JOS. KÖSEL, Publisher, KEMPTEN (Bavaria).

CODICIL

of My Will for Healthy and Sick People

by

Msgr. Sebastian Kneipp,

privy chamberlain to the Pope, parish priest of Woerishofen.

8°. VIII—376 pages. With Frontispiece (prelate Kneipp's legacy to Woerishofen), eight Coloured Plates and numerous Illustrations inserted in the Text.

I. Elegant **Edition.** Price $ 1.65. II. Popular Edition for America. Price in Paper binding, net 50 cts., cloth, net 75 cts.

C o n t e n t s.

Preface. — **I. General.** 1, A word to the sick. 2, Past and present. 3, Something for housekeepers. — **II. Diet.** Well meant and well tested cooking recipes for healthy and sick people. Introduction. 1, Food for the healthy. (Meat. — Vinegar, salt, spices. — Fruit. — Fine pastry and confectionery. — Drinks.) 2, Nourishment for the sick. General remarks. Peritonitis. Abdominal typhus. Acute catarrh of the stomach. Chronic catarrh of the stomach. Inflammation of the stomach. Costiveness. Chronic diarrhoea. Hemorrhoids. Kidney complaints. Diabetes. Complaint of the bladder. Chronic catarrh of the bladder. Poverty of blood and scrofula. Gout. Febrile diseases. The diet of convalescents. Cookery receipts for persons suffering from stomach or feverish complaints, as well as for convalescents. — **III. Gymnastic Exercises** (to be practiced in the room). General. Description of the exercises. Movements of the head. Arm- and hand-exercises. Foot- and leg-exercises. Exercises performed with the bar. — **IV. The Structure and Care of the Human Body.** — Structure of the human body, its functions and the purpose of its organs. General. Division of the body. The noble parts of the body. The heart and the circulation of the blood. The stomach. The intestinal canal. The liver. The organs of digestion. The milt. The brain and the nervous system. The eye or the sense of sight. The sense of hearing. — The care of the human organism. 1, General. 2, The care of the eyes. 3, The care of the ears. 4, The care of the nose. 5, The care of the hair. 6, The care of the teeth. 7, The care of the face. 8, The care of the mouth. 9, The care of the throat. 10, The care of the organs of speech. 11, The nervous system. 12, The care of the lungs. — **V. Something from God's Garden.** 1, Carline thistle. 2, Chickweed. 3, Burdock. 4, Centaury 5, Knot-grass. 6, Pewter-grass. 7, The onion. — **VI. Some Advice for little Accidents.** Practical instruction for immediate help in case of accidents. Introduction. Fractures. The transport of the patient. How to save one's self and others from being drowned. Artificial breathing. Suffocation. Freezing to death. Burns. Swoon. Heat-apoplexy. Poisoning. Hemorrhages. Hernias. Struck by lightning. Dislocations. Sprains — **VII. Diseases.** Disease of the eyes, Egyptian. Complaint of the eyes, syphilitic. Smallpox. Cholera. Epilepsy or falling sickness. Obesity. Fistula. Tetters. Foot complaint. Melancholy and Insanity. Hair and nails. Hysteria. The hooping-cough. The white swelling. The itch. Cancer. Liver complaints. Leanness. Malaria fever. Measles. Milt complaints. Atrophy of the muscles. Nervousuess (nervous disorder). Kidney complaints. Rheumatism. Consumption of the spinal marrow (tabes dorsalis). Delirium tremens. The scurvy. Scarlet. Perspiration. Seasickness. Tonic spasm. Stuttering. Syphilis. Frenzy (hydrophobia). Typhus. St. Vitus' dance. Burn and freezing to death. Teething of children. Diabetes. — **VIII. Various Remarks.** 1, On letting blood (bleeding; cupping glasses). 2, Ice-compress. 3, Vomiting. 4, A word about laxatives and mineral waters. 5, Clay. 6, A chapter on teas.

Verlag der Jos. Kösel'schen Buchhandlung in Kempten.

Foreign Editions of Father Kneipp's Works.

A. German Editions.

Mein Testament für Gesunde und Kranke von Msgr. **Seb. Kneipp**, päpstl. Geheimkämmerer, Pfarrer in Wörishofen. 8°. 25½ Bogen. Mit 29 Vollbildern in Autotypie. 12. Auflage. Preis broch. M. 2.80, gebd. M. 3.40.

Meine Wasser-Kur durch mehr als 35 Jahre erprobt und geschrieben zur Heilung der Krankheiten und Erhaltung der Gesundheit von S e b. K n e i p p. Mit dem autotypischen Bildnisse des Verfassers und vielen in den Text gedruckten Abbildungen. 63. Auflage. 8. X und 376 Seiten. Preis broch. M. 2.60, solid gebunden mit Lederrücken und Goldtitel M. 3.20.

So sollt ihr leben! Winke und Rathschläge für Gesunde und Kranke zu einer einfachen, vernünftigen Lebensweise und einer naturgemässen Heilmethode von S e b. K n e i p p, Pfarrer in Wörishofen (Bayern). 24. Auflage. 8. XII und 364 Seiten. Preis broch. M. 2,60, geb. in R. u. E. Leder mit Goldtitel M. 3.20.

Oeffentliche Vorträge gehalten vor seinen Kurgästen in der Wandelbahn zu Wörishofen von Msgr. **Sebastian Kneipp**, päpstl. Geheimkämmerer, Pfarrer in Wörishofen. E r s t e r B a n d : Die Vorträge des Jahres 1829. Nach stenographischen Aufzeichnungen bearbeitet und herausgegeben von Priester J o h a n n G r u b e r , ehemaliger Privatsekretär von Pfarrer Kneipp. Mit einem Titelbild in Autotypie. 2. Auflage. 8°. 350 Seiten. Preis M. 2.60, gebd. M. 3.20.

Z w e i t e r B a n d : Die Vorträge des Jahres 1894 (April bis September) mit einem Vorworte Kneipp's. In dessen Auftrag gesammelt und herausgegeben von L o u i s e M a r i e S c h w e i z e r. Mit einem Titelbilde in Autotypie. 8°. 344 Seiten. Preis broch. M. 2.60, gebd. M. 3.20.

D r i t t e r Band: Die Vorträge der Jahre 1893/94. Herausgegeben von Prior Fr. B o n i f a z R e i l e , Sekretär des hochw. Herrn Pfarrers Kneipp in Wörishofen, und H. H a r t m a n n. 8°. 336 Seiten. Preis broch. M. 2.60, gebd. M. 3.20.

Ein v i e r t e r Band, enthaltend die Vorträge seit September 1894, wird im Laufe des Jahres 1896 erscheinen.

Pflanzen-Atlas zu S e b. K n e i p p 's S c h r i f t e n , enthaltend die Beschreibung und naturgetreue bildliche Darstellung von sämmtlichen in den Kneipp'schen Büchern besprochenen, sowie noch einigen anderen vom Volke vielgebrauchten Heilpflanzen. **Ausgabe I** (in einfachem Lichtdrucke), 20 Tafeln. 4. Aufl. Preis broch M. 3.60, gebd. in Ganzlwd. M. 5.20. **Ausgabe II** (in Farbenlichtdruck), 41 Tafeln. 6. Aufl. Preis broch. M. 8.—, gbd. in Ganzleinwand M. 10.—. **Ausgabe III** (die Bilder schwarz in Holzschnitt ausgeführt) ohne beschreibenden Text. D r i t t e A u f l a g e. Preis broch. 80 Pf., gebd. M. 1,20.

B. French Editions.

Mon Testament. Avis et conseils aux malades et aux gens bien portants par Msgr. S. Kneipp, curé de Wörishofen. Beau vol. in 12º, orné de gravures, broché Fr. 3,50, relié Fr. 4,25.

COMMENT IL FAUT VIVRE. Avertissements et conseils s'adressant aux malades et aux gens bien portants pour vivre d'après une hygiène simple et raisonnable et une thérapeutique conforme à la nature par M. l'Abbé S. KNEIPP Curé de Wœrishofen (Bavière). Seule traduction française complète autorisée et reconnue authentique par l'auteur, et contenant les gravures des plantes recommandées par M. KNEIPP. BEAU VOLUME petit in-8 (XII-336 pages), orné d'un portrait de l'auteur avec sa signature, et de toutes les gravures de l'original allemand. (Appendice sur les Applications d'Eau) . . . . . 3 fr. 5(

Le même, en élégante reliure . . . . . . 4 fr. 2!

Comment il faut vivre est l'ouvrage de M. KNEIPP qui a obtenu le plus grand succès. Dans l'espace de deux ans 80,000 exemplaires ao sou répandus en Allemagne, et de nombreuses traductions sont déjà publiées ou en préparation. Loin de diminuer le prestige de l'abbé KNEIPP, cet ouvrage lui a vali de nombreux adeptes parmi les medecins. Ceux-là même qui s'étaient défiés de ses remèdes ont reconnu que ses conseils d'hygiène étaient dictés par une haute sage-ai et une connaissance profonde des conditions vitales. Cet homme — les plus récalcitrans sont forcés d'en convenir — est doué d'une puissance d'observation extraordinaire: son regard pénètro plus avant dans le corps humain qu'on ne l'a jamais fait Comment il faut vivre mérite de devenir et deviendra le livre classique de l'hygiène. Il fera époque, croyons-nous, en Allemagne, et dans beaucoup d'autres pays, avec quelques modifications nécessaires peut-être.

SEB. KNEIPP, SOINS A DONNER aux enfants dans l'état de santé et dans l'état de maladie ou conseils sur l'hygiène et la médecine de l'enfan.. Seule traduction française autorisée par l'auteur. Beau volume in-12, orné d'un portrait 2 fr. 00; franco 2 fr. 2!

Le même, en élégante reliure souple . . 2 fr. 75; franco 3 fr. 0(

ALMANACH-KNEIPP pour l'année 1896 rédigé par l'abbé SEBAST' KNEIPP. Traduit en français avec son autorisation exclusive. Beau volume in-18, illustré de nombreuses gravures, broché à 0 fr. 70, car tonné à 0 fr. 95.

MANIÈRE DE PRATIQUER LES APPLICATIONS D'EAU à Wörishofen sous les contrôle de M. l'Abbé KNEIPP. Supplément à l'ouvrage „Comment il faut vivre." In-8, avec figures, 0 fr. 30

UN CURÉ ALLEMAND extraordinaire. Etude sur M. l'Abbé SEB KNEIPP. Par M. l'Abbé A. Kannengiesser. Brochure in-12 avec portrait 0 fr. 75.

LES CURES PITTORESQUES de l'Abbé Kneipp à Wörishofen. Silhouttes et récits d'un touriste par Ernest Goethals. 1 beau vol in-12 de 150 pages et 12 gravures. fr. 2.—

LA MÉTHODE KNEIPP considérations s'adressant à ses partisans et à ses détracteurs aux malades et aux gens bien portants par G. WAGNER. Belle brochure in-12 0 fr. 80.

NEUENS, MANUEL PRATIQUE et raisonnée de l'hydrothérapie de l'Abbé Kneipp. 1 beau vol. in-12 de 160 pages. Prix broché Fr. 1,50.

JOS. KÖSEL, Publisher, KEMPTEN (Bavaria).

Ulsamer, Jean Alfred, PHARMACIE DOMESTIQUE. Recueil des plantes médicinales qui doivent se trouver dans toute pharmacie bien ordonnée. Collection faite pour le peuple dans les jardins, les prairies, les chamos et les forêts. Traduction française autorisée, illustrée de nombreuses gravures. Prix broché Pr. 1.50, en reliure Fr. 1.75.

C. Dutch Editions.

Mijn Testament voor Gezonden en Zieken door Msgr. Seb. Kneipp, Schrijver van „Mijne Waterkuure". Tweede Uitgaaf. Preis broch. M. 2.80, gebd. M. 3.40.

Mijne Waterkuur sinds 30 Jaren toegepast en goed bevonden. Geschreven tot genezing der ziekten en tot behond der gezondheid door Seb. Kneipp naar de 33ste uitgave uit het Duitsch vertaalt. 8. Preis broch. M. 2.80.

Zóó zult gij leven! Wenken en raadgevingen aan gezonden en zieken voor eene eenvoudige, verstandige levenswijze en eene natuurlijke geneesmethode door Sebastian Kneipp, naar de 8ste uitgave uit het Duitsch vertaalt. 8. Preis broch. M. 2.80.

Planten-Atlas tot Seb. Kneipp's Water-Kuur inhoudende de Beschrijving en getrouwe afbeelding volgens de natuur van alle in genoemd boek bosproken en nog eenige andere door het volk veelvuldig gebruikte Genees-planten. Uitgave I (Lichtdruck met tint). Preis broch. *f* 2.20, geb. *f* 3.15. Uitgave II (mit kleurcnlichtdruck). Preis broch. *f* 4.80, gebd. *f* 6.—. Uitgave III (zoart in houdsnee of Chemigraphie). Preis broch. 50 cts., in halb Lwd. gebd. 60 cts.

Raadgever voor Gezonden en Zieken door Sebastian Kneipp. Uit het Duitsch vertaalt door een dankbaren Kurgast. 8. Preis *f* 1.— = M. 1.60.

D. Spanish Editions.

Mi Testamento dedicado á sanos y enfermos por Monseñor Sebastián Kneipp, camarero privado del papa y cura-párroco de Wörishofen (Baviera). Vertido al castellano por D. Joaquin Collet y Gurguí, doctor en medicina por la facultad dc Munich. 8. Mit 29 Vollbildern in Autotypie. Preis in Ganzlwd. geb. M. 4,20.

MÉTODO DE HIDROTERAPIA aplicado durante mas de 35 años y escrito para el tratamiento de los enfermos y para guia de los sanos por Seb. Kneipp, cura Párroco dc Wörishofen (Baviera). Version española de la 33ª edición al emana por D. Francisco de C. Ayuso, catcdrático de esta lcngua. 5. Aufl. Preis broch. M. 3,60, in ganz Lwd. gebd. M. 4.20.

Como habéis de vivir.

Avisos y Consejos para Sanos y Enfermos ó Reglas para vivir conforme à la sana razón y curar las enfermedades según los preceptos de la naturaleza por Sebastián Kneipp, Párroco de Wörishofen (Baviera). Versión espanola por D. Francisco G. de Ayuso. 3. Auflage. Preis broch. M. 3.60, in ganz Leinwand gebd. M. 4.20.

Atlas de las Plantas

usadas en el Sistema Hidroterápico de Kneipp. Edición primera, en negro solamente. Preis broch. M. 4.40, gbd. M. 5.60. Edición segunda, en chromoheliotipia. Preis broch. M. 8.80, gebd. M. 10.80. Edición tercera, en grabados madera. Preis broch. M. 1.—, in ganz Lwd. gebd. M. 1.40.

El cuidado de los niños.

Avisos y consejos para tratarlos en el estado de salud y en la enfermedades por Monsenor Sebastián Kneipp, Camarero privado de SS. León XIII. y cura párroco de Wörishofen (Baviera). Vertido de la sexta edición alemana por D. Francisco G. Ayuso, academico de la espanola 2. Aufl. Preis broch. M. 1.80, gbd. M. 2.30.

Calendario Kneipp

(Almanaque Kneipp) ilustrado con profusión de grabados, redactado por D. Sebastián Kneipp con la cooperacion de varios medicos alemanes para 1894, 1895, 1896, en 8º. Preis broch. jo 80 Pfg.

E. Italian Editions.

IL MIO TESTAMENTO

per sani ed ammalati con 29 Vignette in autotipia. ●duzione autorizzata da Romeo Lovera. Preis broch. M. 2.80, gebd. M. 3.40.

La mia cura idroterapica

esperimentata per oltre 35 anni e scritta pella guarigione delle malattie e pel mantenimento della salute da Seb. Kneipp, Parroco in Wörishofen (Baviera). Col ritratto dell' Autore. Traduzione antorizzata fatta da L. Moltini e dal Dr. O. de Guggenberg. 4. Aufl. Preis broch. M. 2.80, gbd. M. 3.40.

Cosi dovete vivere!

Avvertimenti e consigli dati ai sani ed agli ammalati per vivere nella maniera più semplice e naturale coll'aggiunta d'un metodo di cura naturale ver Sebastiano Kneipp, Parroco in Wörishofen in Bavaria. 3. Auflage. Preis broch. M. 2.80, gbd. M. 3.40.

La cura dei bambini

sani ed ammalati. Consigli del Parroco Sebastiano Kneipp, cameriere segreto di S. S. il Papa. Traduzione autorizzata, fatta sulla VII. edizione tedesca da Lorenzo Moltini. 8. 168 pp. Preis broch. M. 1.60, gebd. M. 2.10